ROMAN SCIENCE

WILLIAM H. STAHL

ROMAN SCIENCE

ORIGINS, DEVELOPMENT,

AND INFLUENCE

TO THE LATER MIDDLE AGES

MADISON 1962

THE UNIVERSITY OF WISCONSIN PRESS

Published by
The University of Wisconsin Press
430 Sterling Court, Madison 6, Wisconsin

Copyright © 1962 by
the Regents of the University of Wisconsin

Printed in the United States of America
by North Central Publishing Co., St. Paul, Minnesota

Library of Congress Catalog Card Number 62-9263

Preface

The present study had its origin in a paper prepared for a symposium of the History of Science Society on classical influences upon the science of the Middle Ages. To keep the paper within reasonable limits, attention was focused upon three postclassical Latin writers who stood on the threshold of the Middle Ages. The choice of Chalcidius, Macrobius, and Martianus Capella proved to be a happy one because all three had a far-reaching influence upon medieval science and all were purveying doctrines that originated in the Hellenistic and earlier ages in Greece. Two questions continued to occupy me after completing the paper: Would an examination of the works of other Latin writers on science bear out the findings; and if the evidence continued to point to two men—Varro and Posidonius—as the chief transmitters of Greek science to Rome, what were the origins of the materials they were transmitting? In seeking to answer these questions it became evident that what was being undertaken was a book on Roman science, looking backward to its origins and forward to its influence.

The book is addressed to readers who are interested in the intellectual history of Western Europe. It may be of particular interest (1) to students of the history of science, because, to my knowledge, this is the first separate study of the development of Roman science; (2) to students of the classics, since what it describes is the science that was known to educated laymen, Cicero, Lucretius, and Virgil among them; and (3) to students of medieval civilization, since the purviews and practices of Roman writers on science and philosophy had a controlling influence upon the literature in those fields in the Latin West for over a thousand years.

Questions will naturally arise in readers' minds about the appropriateness of the broad title given to this small book. If it may rightly lay claim to being the first book on the subject, there is really nothing remarkable about such a claim; what is remarkable is that it may be said to be a book on a nonexistent subject. The matters given primary attention here cannot strictly be classified as either science or Roman. The fact remains, however, that "Roman science" is a term that is widely used, and misapprehensions about its nature still exist. If this volume contributes in a small measure to clearing up some of those misapprehensions, it will have served its purpose. Objections to the title may also be raised because the volume is devoted chiefly to the subjects of the quadrivium —arithmetic, geometry (and geography), astronomy, and harmonic theory. It should be pointed out here that the purpose of the book has not been to offer a comprehensive account of Roman science but to illuminate the character of the scientific literature by tracing some aspects of Latin science from their origins to their vestiges in the later Middle Ages. The science of the Romans, such as it was, was textbook science; and the extant Latin texts show the lines of transmission more clearly in quadrivium subjects than in others. A few subjects outside the quadrivium have been discussed, but only to show that they conform to familiar patterns. The conclusions presented here, the writer is convinced, would apply as readily to other aspects of Roman nontechnological science. Admittedly, then, a more accurate title for this volume would have been: "Roman Scientific Compilations; an Essay on Some Aspects of Roman Nontechnological Scientific Interests."

The Fulbright authorities provided me the leisure time to begin reading and reflecting on these questions, for the invitation to prepare the original paper came while I was serving as Fulbright lecturer at the University of Melbourne in 1956. I am deeply grateful to Carl Boyer of Brooklyn College, I. E. Drabkin of the City College of New York, Ludwig Edelstein of the Rockefeller Institute, and Marshall Clagett and W. D. Stahlman of the University of Wisconsin for their many helpful criticisms of my manuscript; and to Otto Neugebauer of Brown University for offering corrections of several points in the preliminary paper as well as drawing my attention to a seldom-recorded scheme for rising times in Martianus Capella's encyclopedia. I also wish to thank Bluma Trell of New York University and Ethyle Wolfe of Brooklyn College for valuable comments and suggestions. Material aid was provided by the staffs of the libraries of Brooklyn College and University College of New York University in procuring books not easily available. I owe much to Roberta Yerkes for the care and acumen that she brought to bear in readying the manuscript for press. Lastly I wish to record a special debt of gratitude to my wife for her encouragement and for many hours of patient collaboration, in assisting with the reading of the manuscript and offering her reactions to my phraseology at every stage.

W. H. STAHL

December 1961

Contents

PART THREE
ROMAN SCIENCE IN THE MIDDLE AGES

PART ONE

GREEK ORIGINS

I

INTRODUCTION

Roman science has been a much neglected field of study. Because it can be reasonably asked not only whether it was really science but whether it was Roman, many authorities avoid the second question and warily refer to "Greco-Roman science." In that it was essentially derivative in character, Roman science presents an analogy to Arabic science. The striking difference lies in the fact that whereas Arabic science was of a rather high order, being derived from the finest products of Greek scientific genius, Roman science was of a distinctly low order. Arabic science has always been a fertile field of investigation, to professional scientists as well as to historians, even though it has been generally understood that the Arabs themselves contributed very little to their science. Roman science, on the other hand, has attracted few admirers and has received cautious or even patronizing handling from historians of science—a deference accorded as fitting regard for one aspect of the civilization of a dynamic people who cut a broad swath in history.

Anyone who studies the body of Roman jurisprudence is predisposed to marvel at its systematic character and its scope, aware that the Romans laid the foundations of this profession and were pre-eminent in its development. But when an admirer of the ancient Greeks and Romans reads a chapter on Roman science, he approaches the subject with a wistful attitude, expecting to find little of merit and hoping that the author will turn up some unsuspected facets which will serve to give the Romans a decent rating here and make it appear that they must have had a fairly rounded development. To give deferential or sympathetic treatment to a particular phase of intellectual history simply because it is representative of a dominant civilization may be a grievous mistake.

Historians of science, aware of the hopes and expectations of their more solicitous readers, have been careful in the selection of materials for their accounts of Roman science. We usually find, in the larger works on early science, between the major parts devoted to Greek and Byzantine or Arabic science, a section on the Romans, expanded to a size intended to reflect their greatness as a nation. This section generally consists of digests of the scientific portions of outstanding works, together with examples of remarkable scientific observations recorded, but not originated, by Roman writers. But what credit is it to the Romans if the first extant mention of a scientific discovery or observation was made by a Latin author whose entire work was derived from Greek sources? If the earliest example of an astrolabe happened to be found in a Roman ruin, we would not suppose it was a product of Roman ingenuity.

It has been a common practice, when dealing with Lucretius, to emphasize the strikingly modern features of his atomic theory, so that the reader is often left with the impression that Lucretius was the precursor and antecedent of Dalton. Little is said of the childishness of many of Lucretius' views about the physical world and of the fact that Epicurus, the font and origin of his doctrines, was himself, as a scientific philosopher, often centuries behind his time. Care is taken to cull the modern observations and techniques in other larger repositories of Roman scientific information—Pliny's *Natural History* and Seneca's *Natural Questions*—and little insight is afforded into the works as a whole. Pliny's chapters on mineralogy and metallurgy, for example, contain many statements that evoke the plaudits of experts in those fields. When Gibbon aptly described the *Natural History* as "that immense register where Pliny deposited the discoveries, the arts, and the errors of mankind,"[1] he had an appreciation of the character of the work as a whole; but historians of science tend to emphasize the discoveries and the arts and to pass lightly over the errors. To round out the section on Roman science, some briefer accounts are generally included of Celsus' *On*

Medicine, Vitruvius' *On Architecture,* Mela's *Chorography,* Frontinus' *On Aqueducts,* and Cato's *On Farming,* without consideration of whether the material is Roman and how much of it is science. Some writers, disheartened by the dearth of scientific matter, arrange their topics under the heading "Roman Period" and thus are able to introduce the two imposing Alexandrian figures, Ptolemy and Galen, to give bulk to the Roman section. Readers sometimes get a stronger impression of Galen as a Roman court physician to the emperor Marcus Aurelius than as a Greek researcher at Alexandria, Pergamum, and Smyrna. Is it any wonder that casual readers are confused as to what Roman science really was?

Another frequent resort to enable writers to present Roman science in a favorable light is to draw attention to extraordinary Roman achievements in the fields of engineering and the mechanical arts. Such discussions properly belong in a history of technology and not in a history of science. Pliny's mineralogy and metallurgy, Vitruvius' *On Architecture,* and Frontinus' *On Aqueducts* are cases in point, even though Frontinus' treatise demonstrates skill in calculation and an understanding of the geometrical manipulations involved in his operations, and though Vitruvius' recommended course of training for would-be architects includes geometry (to understand the use of the rule, compasses, set square, and level) and arithmetic (to reckon building costs) among various subjects that have usefulness to the architect, such as history, philosophy, music, medicine, and astronomy.

The Romans undoubtedly produced better engineers than the Greeks; but Roman engineers had no scientific training, and the small amount of mathematics they required could be applied in practice without any grasp of theory. Even among the Greeks the proportion of the population engaged in scientific research or exposed to scientific training was minute. Greek mathematics and astronomy were rarely applied to problems of engineering; and the Greeks seldom brought their scientific knowledge to bear upon practical matters—save in the arts of warfare,[2] where the motivation was evidently sufficient. Many of the celebrated devices of their mathematical mechanicians were used solely to demonstrate theories. Among the few instances that can be pointed out of the actual application of Greek theory to practical subjects, two are related to developments discussed in this volume: the theoretical treatises on sundials and on mathematical geography.

The difficulties in attempting to ascertain the extent of Roman indebtedness to the Greeks in technological fields are immeasurably greater than in tracing the lines of transmission of theoretical science from Greece to Rome. Theoreticians are intellectuals who put their thoughts in writing and leave the marks of their indebtedness; but the engineers and mechanics of antiquity were not concerned about how a technique, device, or implement in daily use originated.[3] We know however that the Romans frequently employed Greek technicians in such complicated undertakings as the construction of siege engines or aqueducts.

It should not be assumed that skill in mechanics and the arts and crafts among the peoples of the ancient world reflects creditably upon their intellectual capacities. Ancient technicians and scientists for the most part lived in worlds separate from each other. Some scientific speculations, such as man's theorizing about the causes of disease or his principles for preserving good health, or his conceptions of the geographical and cosmographical world about him, serve as fairly reliable indexes of the level of his intellectual attainments and sophistication. The history of technology, by contrast, shows little if any conformity to the patterns of gradual evolution, retardation, and reawakening that we are accustomed to observe in the intellectual history of early times, the Middle Ages, and the Renaissance. A visit to a large museum will remind us that many arts, crafts, and techniques reached the pinnacle of their development in ages long past. The manual dexterity and some of the basic tools of primitive savages could scarcely be improved upon. During the later Middle Ages scientists and craftsmen did begin to join hands in their search for the answers to questions, and some of the revolutionary advances of technology resulted from the changed attitudes of scientists who were no longer disdainful of the occupations of technicians. But, in general, the high level of medieval technology presents a striking contrast with the low level of intellectual life. While miners were driving shafts deep into the earth, shoring them up with underpinnings of massive beams, and draining water from the floors of the mines with piston suction pumps or elaborate sets of chain buckets, such as are described in Agricola's *De re metallica*, their brothers in a nearby monastery or school were engaged in copying manuscripts that they could not comprehend.

This book makes no pretense of presenting a balanced and inclusive account of Roman scientific activities and applications. A well-proportioned study of Roman medicine, for instance, would have to focus major attention upon Roman successes in organizing public health and military medical services and upon Roman developments in public sanitation and hygiene, subjects that are beyond the scope of the present study, which is restricted to science that was transmitted through books. It is hoped, however, that the present study, by dwelling upon the derivation of Roman science from lay Greek sources, will counterbalance some of the effects of historians who have looked for appreciative things to say about Roman science. It has been felt necessary to digest the contents of important authors like Lucretius, Celsus, and Pliny; most of the major writers of the classical period have already been translated. Readers are urged to study their works rather than to depend upon digests which, by their very selectivity, nearly always reflect personal attitudes. Some later writers have been digested here (again, perhaps, not without bias) because translations have not yet become available. Lastly, it must be acknowledged, any study that endeavors to provide new insights into the true character of a subject by drawing particular attention to how completely and ineptly it was derived from earlier sources inevitably presents that subject in its worst light.

The Greeks dealt with science on two divergent levels—the truly scientific level of original creative research and the popular level of lay writers of handbooks and commentaries. Greek creative science continued to flourish throughout classical antiquity, long after other aspects of Greek intellectual life had shown a decline. Astronomy, geography, and medicine reached the culmination of their development in the ancient world in the works of Ptolemy and Galen, who lived in the second century of the Christian Era; and some of the most brilliant Greek mathematicians—Diophantus, Pappus, Theon of Alexandria, and Hypatia—lived in the third and four centuries. Scientific studies, fostered and maintained on this high level at Alexandria and Byzantium, were transmitted directly, or through Syriac translations, to the Arabic world.

At the top of the scientific world were such luminaries as Hippocrates, Democritus, Eudoxus, Euclid, Aristarchus of Samos, Archimedes, Eratosthenes, Apollonius of Perga, Hipparchus, Herophilus, Erasistratus, Ptolemy, and Galen, men of extraordinary talents, who succeeded in making their age one of the most significant in the history of science. Their teachers were scientists, and their proficiencies were further developed by studying the works of their predecessors. If they happened to live contemporaneously, they were usually in touch with each other, though they might be separated by a thousand or more miles; they showed their esteem by dedicating works to each other. Their technical treatises had a very limited audience—one of the chief reasons that few original scientific writings have survived from antiquity—and their circle of intimates was small. But as long as original Greek science flourished, popularizers communicated some of the results to the educated public. The existence of the popularizers, as we might expect, was anything but secluded. Their books enjoyed wide circulation and they often achieved greater reputations than the men whose works they were explicating or digesting. During the Hellenistic Age the trend of popularization kept growing, for reasons which will be explained later, and as a result even leading scientists were occasionally drawn into writing books for educated laymen.

Let us consider a few instances. Eratosthenes, the founder of mathematical geography and a brilliant figure in the history of Greek mathematics, was better known to his contemporaries as a literary man. His application of his specialized skills to the interpretation of works of Homer and Plato attracted more attention than his original scientific discoveries. The case of Hipparchus is an ironical one. He was the greatest astronomer in antiquity, yet his fame in his own day was overshadowed by that of a poet, Aratus, who had composed a didactic poem on the constellations to clarify allusions in classical poetry. Because of the vogue of the poem, Hipparchus wrote a commentary on it, undertaking to correct its errors. Of all the truly remarkable researches and publications credited to Hipparchus, the one work to be preserved was this commentary, its fortunes determined not by its intrinsic merits, which were considerable, but by the popularity of a poem. Ptolemy was a scientist who catered occasionally to popular tastes. His large work on astrology, in four books, was until recently the only one of his works that was available in English translation. Thus it has often been. The original thinker has been left to commune with his colleagues; the popularizer, the idol of the public, has also been the darling of the publishing profession.

Even Romans who had the keenest scientific aptitudes and interests never went more deeply into the study of Greek science than an intelligent layman would regard as necessary for an understanding of literature and natural philosophy. We also find, as we trace the sources of Roman science, that most Greek intellectuals had the same attitude. Science was to them, too, a generally unattractive field, not to be compared with the delights of literature and philosophy; but a certain amount of basic schooling in scientific subjects was admittedly essential to comprehend the more abstruse doctrines of the philosophers or to demonstrate one's acumen in disclosing hidden or symbolic meanings in the works of the classical poets, meanings that the poets, to be sure, never intended to convey.

This lay interest in an elementary knowledge of science is observed throughout the classical Greek age and may be traced back as far as the early Pythagoreans. They were the first Greek intellectuals to provide their disciples with a broad general curriculum emphasizing the key position of mathematical studies. In fact the very subjects that were to become the set program of the mathematical quadrivium in the Middle Ages were recommended for the Pythagorean curriculum by Archytas of Tarentum (c. 380 B.C.).[4] Some of the early Pythagoreans were polymaths and, in drawing their students away from specialization to the cultivation of a general knowledge of many subjects, were laying the foundations for the liberal arts program of studies that has had the support of Western intellectuals for over two thousand years. Then, in the late fifth and early fourth centuries, the philosophers and rhetoricians who were called the Sophists became interested in general education, in their case to train gifted young men for political leadership. The Sophists disagreed about the importance of mathematics in the curriculum, but even those who were opposed to giving it much attention admitted that it had value as a disciplinary subject. Plato strengthened the position of mathematical studies in the schools by insisting, in the seventh book of the *Republic*, that students being trained to become leaders of the state have thorough instruction in arithmetic, geometry, stereometry, and astronomy. It is ironic to observe Greek efforts to promote the study of the liberal arts as preparation for political leadership at the very time that they were losing their political freedom. Aristotle, through his studies of many fields of knowledge and systematization of them, became a powerful influence in the trend toward polymathy among the Greeks; but it was his successors, with their bent for writing textbooks in the separate fields of knowledge, who gave the handbook movement its full momentum.

After the subjugation of their country by Philip, king of Macedon, and his son Alexander, Greek intellectuals lost most of their zest for politics and became introspective, taking up the study of philosophy as compensation for their loss of outward glory. Soon it was philosophy that provided motivation for pursuing liberal arts studies, whether a young man aspired to become a diligent disciple of a philosophical school or merely wished to be able to follow recondite philosophical disquisitions with understanding.

About the same time, literary creativeness declined sharply in Greece. The classical poets and masters of prose had run the gamut of the literary genres, developing each to standards of excellence that frustrated emulators. There was little to do but follow the example of the learned critics of the Alexandrian Library who contemplated the past, classified works into canons, and exercised their skill in interpreting the thoughts of the masters. These erudites seldom evinced appreciation of the literary qualities of the classical poets and prose stylists. To them the poets were merely men of prodigious learning, experts in all fields of knowledge. The majestic surge of Homer's verse, the vivacity of his narrative episodes, and the gusto of his heroes were lost upon explicators who were intent upon making a pedant of him. Homer and Aeschylus were to be admired for their knowledge of world geography, for example, and it required a commentator's acumen and lucubrations to point out how cleverly the poets had veiled their learning in their verse.

Such a combination of reverence for the work under explication and zeal to impress readers with the range of the commentator's learning inevitably resulted in ascribing to an early poet or thinker knowledge that was not discovered until centuries after his death. Homer was interpreted as if he had been acquainted with late Hellenistic geography, Plato as an adept in late Hellenistic astronomy. Commentators on Homer soon began to draw their admiring pupils away from reading his poems, and commentaries on the *Timaeus* supplanted the original treatise in Platonist schools. Literary genius had virtually expired and the learned pedant had taken over. For such artful and tortuous displays of a commentator's learning in varied fields of knowledge a course of liberal arts studies provided ideal preparation.

It was not until the first half of the second century B.C. that the Romans were able to gain extensive contacts with polite Greek society;

that significant phase in intellectual history began as Roman armies were conquering the Greek states. By that time cultured Greeks had settled back to contemplating the greatness of the past and enhancing their appreciation of the subtleties of the literary classics by acquiring broad philosophical and scholarly backgrounds. The Romans had never shown any inclination or aptitude for intensive theoretical studies; the Greeks were now beginning to be cloyed with their dilettantism and sophistication. Thus the mingling of Greek and Roman cultures took place under circumstances congenial to both societies. Roman senators and wealthy aristocrats quickly conceived a passion for dipping into Greek learning, and Greek intellectuals felt admiration for the dynamic qualities of a nation that had produced such efficient armies and demonstrated such talent for administrative organization and public works as was evident in the empire that the Romans were building.

It has been frequently pointed out that the Romans, when they were first initiated into studying Greek philosophy, took to some of its teachings as enthusiastically as if they had been produced by Roman minds —to Stoicism in particular, with its doctrines about pantheism, divine intervention, morality, sense of duty, and insensitiveness to pleasure or pain. Furthermore, it appears that the attractive, intelligible, eclectic forms of philosophy that were being developed in Greece in the second and first centuries B.C. and which Roman intellectuals found so congenial to their tastes were a response to a yearning of the Greeks to find inner satisfaction in an unchallenging world. The trends that were responsible for shaping the later Hellenistic philosophies were not unlike those that induced large numbers of people to seek various kinds of knowledge in handbooks.

Greek compilers of the later Hellenistic period, responding to a current fad for acquiring bits of information on every subject including the most trivial, had produced a spate of miscellanies and specialized compendia, some of them on technical subjects. There were treatises on the various operations performed by farmers and soldiers, on precious gems, gastronomy, fishing, and even a book on cosmetics, ascribed, not surprisingly, to that eminent authority on the subject, Cleopatra. Once the Romans had acquired a liking for handbooks, they showed no inhibitions about enlarging the scope of their applications. Having no sense of the dignity of the writer's profession, they undertook to compile manuals on the most mundane subjects. Where earlier works on a subject were available, as in the case of Carthaginian and Greek treatises on agriculture, Roman writers made extensive use of them. But in practical fields they depended heavily upon their own experience and handled their subjects with self-assurance, sometimes interjecting vigorous criticism of their authorities for not putting their recipes to adequate tests. Roman treatises on practical subjects were often better than any that had previously been produced.

The Romans were a pragmatic and utilitarian people. They valued books for the practical information contained in them. They approved of philosophical studies because they seemed to have application to the moralities of everyday living. And their penchant for straightforward how-to-do-it books far exceeded any such inclination among the Greeks. Students of Roman literature who are familiar with the bulky *Römische Literaturgeschichten* produced in Germany in the nineteenth century are aware that a large proportion of the writings do not strictly belong in the realm of literature. Much of what has survived or is known to have existed is of the informational handbook variety, treatises on every conceivable aspect of daily life: cooking, divination, casting nativities, veterinary medicine, cosmetology, ballistic engines, to mention but a few topics. The Romans were delighted to discover that the Greeks had already compressed the subject matter of various fields of learning into compendia that could be readily translated or further digested for Roman readers. The Greek *technē* or technical treatise became Latin *ars*, the Greek *eisagogē* or *hyphēgēsis* (elementary treatise) became Latin *introductio*, and the Greek *encheiridion* or handbook became Latin *manuale*.

Didactic authors seek to win the approval of readers by impressing them with the authoritativeness of the sources of information being presented in their works. It is quite understandable that venerated lawgivers and men with reputations for wisdom, like Moses, Solomon, Lycurgus, Solon, and Numa, should command respect as authorities in matters of law, religion, and statecraft, for as long a time as their national institutions endured. Early poets, too, like David and Homer, were venerated for their wisdom. So great was the admiration for these men in later ages that coteries of anonymous imitators were happy to have their writings and reflections ascribed to their masters. But in the fields of science and natural philosophy, where views are revised from generation to generation and soon become outmoded, it is curious to find

Greek writers attaching as much importance to early authorities as to later ones. Pythagoras and the early Ionian philosophers were esteemed alongside Aristotle, Archimedes, and Ptolemy. Discrimination was seldom made between primitive and sophisticated concepts, early and late. Intricate sets of epicycles and eccentrics, devised to account for planetary anomalies, are attributed to Pythagoras and Plato, who lived long before the epicycle and eccentric were developed. The uncanny knowledge of the Pharaonic Egyptians and the Chaldeans—whatever is meant by the ancient use of the latter designation—was considered particularly profound in calculation and astronomy. Much of the veneration of the Alexandrian Ptolemy during the Middle Ages stems from his being confused with early Pharaonic astronomers. Even today some mystical religious orders and occult societies claim to be in possession of those potent Eastern secrets, preserved without impairment through thousands of years.

If Greek writers were naive in attributing to men like Pythagoras and Plato views and theories that belong to much later ages, they were also fairly honorable in assigning credit to their later authorities. Not so the Romans of the classical age, who tried to conceal their pillaging of the Greek handbooks by citing early authorities and suppressing the names of the writers they were actually translating or paraphrasing. This practice may be assumed to have prevailed among all Roman authors of didactic treatises, and they passed it on to medieval Latin writers. Some medieval compilers were distinctly more honorable than others; but all the Latin writers of the early Middle Ages referred to in this book were involved in this practice. The first writers to my knowledge who were fairly scrupulous about citing authorities were the encyclopedist Vincent of Beauvais (c. 1190–1263) and some his contemporaries. We must be wary about putting reliance upon the citations of Latin writers before the thirteenth century.

That is not to say that scholarly compilers after the thirteenth century became scrupulous. Free borrowing habits have been common enough in subsequent centuries. Renaissance erudites like Rabelais and Ben Jonson extracted much of their impressive scholarship from handy collections. Textbook writers today often go little farther afield in their researches than the competitive textbooks they are hoping to supersede. Contributors of ostensibly new articles to reputable encyclopedias have occasionally been found indulging in open plagiarism. Many of the masterly original articles on subjects in the humanities have been carried over from nineteenth-century editions with little more than editorial changes.[5] Drastic revisions have often merely entailed drastic reductions, so that earlier editions have had to be kept available on library reference shelves. In subsequent editions editorial consultants have shared equal credit with the authors of the original articles, and occasionally the editor's initials have supplanted those of the author.

Let us examine more closely the course of development of the Greek handbook movement, which was to have as one of its offshoots the bulk of the Latin textbook science transmitted in Western Europe for more than a thousand years. As we do so, we must bear in mind that the Greek handbooks which provided Romans with a smattering of scientific knowledge were generally compiled by laymen and are not to be confused with technical and specialized treatises produced by expert scientists like Euclid, Archimedes, Heron of Alexandria, and Ptolemy. Roman readers did not make the effort to comprehend the matter contained in such professional treatises.

II

CLASSICAL GREEK ORIGINS

It will be readily understood that tracing the origins and development of lay handbook science among the Greeks is quite different from outlining the more significant achievements of Greek science. Most handbooks were intended for the general reader. Such books were occasionally written by men who had made original contributions to Greek scientific thought; more often the authors themselves had limited competence in handling scientific subjects or were laymen used to presenting any subject in a simplified form. Popularizers accumulated materials from random sources with little regard for consistency. We find in some lay accounts opinions of pre-Socratic philosophers intermingled with the theories of Hellenistic scientists who lived three or four centuries later. The framework of a system more often than not was of early origin, because early views tended to be simpler and to serve more conveniently as the basis of a structure. Furthermore, lay compilers showed no preference for the opinions of expert scientists but embraced as readily the likely explanations of literary critics and poets. Scientific systems were dismantled; parts that were complicated or seemed improbable were excluded and the simple and likely parts were incorporated. Heraclides' planetary theory was partially adopted (i.e. heliocentric motions for Venus and Mercury were accepted, diurnal rotation of the earth on its axis was omitted); but the complete heliocentric theory of Aristarchus of Samos, which included diurnal rotation for the earth, was hardly ever alluded to by handbook writers. This fact helps to explain why the complete theory of Aristarchus expired so quickly, and is one of many examples of the important influence that popular science had upon intellectual developments in Greece.

The gathering of materials that lent themselves to use in handbooks began with Pythagoras and the early Pythagoreans. It is true that occasionally views of early Ionian philosophers who antedated Pythagoras are found cropping up in the tradition after Aristotle and his successor Theophrastus gave them currency, but it was the Pythagoreans who inspired the handbook movement and gave it much of its substance and body. It would be hard to overemphasize the importance of the contributions of the early Pythagoreans to that movement and the large share that their doctrines had in its traditions throughout the course of its development to the later Middle Ages. Even in the twelfth century, when lay handbook science began to yield to the obvious superiority of the genuine Greek science that was being introduced into Western Europe from Arabic sources, Pythagoreanism still attracted attention, and this continued into the Renaissance and modern times.[1]

Pythagoreanism introduced two important concepts into the intellectual life of ancient Greece: the desirability of a broad general education and emphasis upon the key position of mathematical studies in that education. The religio-philosophical brotherhood founded by Pythagoras at Croton in southern Italy around 530 B.C. soon developed into a full-fledged educational institution, operating under regulations and holding stated sessions in buildings belonging to the sect. Pythagoras' school is believed to have served as an example and model for the institution of later philosophical schools—Plato's Academy, Aristotle's Lyceum, and Epicurus' Garden.[2] At first instruction was exclusively oral; not until the time of Philolaus, about a century after the founding, were the teachings of the sect committed to writing. As a consequence, it is a matter of conjecture whether doctrines attributed to the master really belong to him or to his early disciples. Because of the reverence in which Pythagoras was held, his followers would have been quite willing to have their discoveries and views credited to him.

Traditions of the school, both early and late, agree in representing

Pythagoras as a man of broad, almost universal, interests. The philosopher Heraclitus, by way of showing his disrespect, lumped him with other polymaths.[3] Even at that early time polymathy was getting a derogatory connotation. Pythagoras' indoctrination of his followers embraced all aspects of their lives. This is not the place to discuss the multiplicity of subjects in the broad course of Pythagorean training, which included even rigid and precise dietary regimens. It is to the point to consider the universality of Pythagoras as a man of science and to refer briefly to those Pythagorean teachings that were to become part of the handbook tradition.

Ancient reports, and some modern scholars, credit the master himself with maintaining the sphericity of the earth and universe, with observing that the planets have independent eastward motions contrary to the motion of the celestial sphere, and even with identifying the Morning and Evening stars as one and the same body. Early Pythagoreans placed the earth at the center of the heavens and the circular paths of the planets at varying distances from the earth. According to a very popular tradition, Pythagoras made the remarkable discovery that numerical ratios underlie the consonant chords. He is reported to have observed that lyre strings varying in length by 2:1, when vibrated, produce the octave, by 3:2 the fifth, and 4:3 the fourth. Pythagoras may then have concluded that if numbers are the key to harmony, they are likely to be the key to the entire universe. Perhaps a connection was drawn between the seven notes of the heptachord and the seven visible planets (the sun and moon were counted among the planets). At any rate, early Pythagoreans believed that the distances of the planets from the earth and from each other were ordained by the numerical ratios present in the harmonious chords. Hence arose the doctrine of the harmony of the spheres.[4]

The discovery of the relationship between music and numbers provided constant encouragement to Pythagoreans in their efforts to demonstrate that all things were numbers. It was the Pythagoreans who gave the word "mathematics" its accepted meaning; previously *mathemata* simply meant "things learned or known." They virtually established the subject as a science. Before their time, with a possible exception in the case of Thales (c. 636–546 B.C.), figures and numbers had been applied merely to surveying and measuring land and to practical reckoning. The Pythagoreans dealt with fundamental concepts of mathematics

—points, lines, surfaces, and bodies—as pure abstractions. The theory of figured numbers may have been originated by Pythagoras. His arithmetical computations were performed not by numerals but by pebbles, counters, or dots placed in the sand. He arranged pebbles in basic geometric forms and observed the relationship between numbers and the shapes of things. A point or dot represented the number one; two dots represented two, and stood for a straight line; three dots represented three and the first plane figure; four dots represented four and the first solid figure. The distinction between odd and even numbers was also traced back to Pythagoras.

Early Pythagorean mathematics provided the bulk of the mathematics of the lay handbook tradition. Commentators and compilers grasped any reasonable opportunity—such as the explanation of the use of numbers by the Divine Creator in fabricating the World-Soul, in Plato's *Timaeus*—to introduce readers to the elements of Pythagorean arithmetic and geometry. Mathematical textbooks also owed a great deal to the early Pythagoreans. The substance of the early books of Euclid's *Elements* was derived from them; and when arithmetic began, in the first two centuries of the Christian Era, to rival geometry in popularity as a mathematical study, particularly in the Latin West, the leading textbooks, such as Nicomachus' *Introduction to Arithmetic*, were largely reproduced from early Pythagorean mathematics.[5] Archytas of Tarentum, referred to earlier as the first Pythagorean to specify the four subjects of the mathematical quadrivium, worked out the numerical ratios of the intervals of the tetrachord on three scales, the enharmonic, the chromatic, and the diatonic, and explained that sound originates from impact, the higher tones resulting from quicker motion being transmitted to the air and the lower tones from slower motion. These doctrines, too, became commonplaces in the handbook tradition. The early Pythagoreans were responsible for the adoption of musical theory as one of the subjects of the mathematical quadrivium, an attitude that was to endure throughout the Middle Ages.

Before leaving the Pythagoreans, attention should be drawn to the daring speculations of some early members of that school in their efforts to explain celestial phenomena. Hicetas and Ecphantus, both citizens of Syracuse, are reported to have maintained that the diurnal rotation of the heavens was only apparent motion and that actually the earth was rotating on its axis. Philolaus (c. 440 B.C.) propounded a system of the heavens that actually made the earth a planet, giving it an orbital motion about a central fire. All three Pythagoreans are cited by Copernicus as precursors of heliocentricism in the prefatory letter to his monumental *De revolutionibus orbium caelestium*, although he suppressed from the body of his work a statement indicating that he was also familiar with the truly heliocentric system of Aristarchus of Samos (c. 310–230 B.C.).[6] The speculations of the three Pythagoreans were too unconventional to receive much notice in the handbooks, but they inspired later theories which became acceptable. The views of Hicetas, Ecphantus, and Philolaus were made known to the Aegean Greek world through the writings of Aristotle and Theophrastus; and since Heraclides of Pontus, whose geoheliocentric ideas were admitted into the handbooks, once studied at Aristotle's school, it is reasonable to suppose that he drew his inspiration from those Pythagorean speculations.

Ever since the collapse of tyrannies in most Greek cities in the sixth century and their replacement by more democratic forms of government, the Greeks had been concerned with how to make democracies workable and, in particular, with developing political leaders who had political virtues, whatever those might be. The basic question had been how best to train a man in political virtue. By the latter half of the fifth century the class of itinerant lecturers and teachers known as Sophists was emerging, who professed to be able to inculcate political ideals in youth. Naturally they attracted much attention to themselves as they argued the merits of their respective systems of education. Most Sophists claimed to be able to teach a multiplicity of subjects. A main point of difference among them was whether mathematical subjects ought to be included in the course of instruction. In the beginning the resistance to mathematics was great, its connection with politics and morality not being very evident. The attitude of most people, including some Sophists, was that the study of mathematics was useless, suited only for the attention of the idle rich. It is to the credit of the Sophists that they finally persuaded themselves of the value of mathematics and gained for it a place in the educational curriculum of the Greeks. In so doing, they deserve to be regarded, as they usually are, as the true founders of the liberal arts tradition in the Western world.[7] The resemblances between the Pythagorean and Sophist movements—the enthusiasm of both for polymathy, mathematical, and abstract studies—are not accidental. Most of the Sophists who introduced the study of

geometry into Athens came from Sicily and presumably got their training from Pythagorean teachers.

Hippias of Elis (c. 481–411) offers the best example of the Sophistic
ideal of a polymath; and in fact may be described as a truly universal
man. In addition to having mastered all known fields of written knowledge, he was skilled in the technical arts as well, and attended a festival
at Olympia wearing jewelry, garments, and embroidered girdle fash-
ioned with his own hands. Hippias was depreciated in antiquity (as he
has also been by modern scholars) as a prime example of the super-
ficiality of polymathy, and is treated by Plato, in two dialogues that
bear his name, as a pompous fool; yet he made a very important contribution to mathematics. He is credited by historians with the brilliant
invention of the quadratrix, the first curve to be recognized by Greek
geometers after the circle. This curve was originally devised to solve
the problem of trisecting any angle, but a century later it was also used
for squaring the circle; hence its name. Hippias took a leading part
in the struggle to gain for mathematics an approved place in the educational program of his day and he deserves to be ranked as one of
the most significant figures in the history of the liberal arts tradition.
Protagoras (c. 485–415), one of the earliest and most renowned of the
Sophists, disapproved of subjecting youth to the tedium of studying
calculation, astronomy, geometry, and musical theory, as we learn from
a dialogue of Plato (*Protagoras* 318d–e), and took issue with Hippias
on this point. But Theodorus, Protagoras' pupil, who was also Plato's
teacher in mathematics, is portrayed by Plato in the *Theaetetus* as a
master of mathematical subjects.

It is with the Sophists that we first clearly discern the cleavage that
was to continue to exist in Greece between original research, the search
for truth (exemplified by the Ionian physicists of the sixth century),
and the teaching of science or the presentation of results in assimilable
form. The Sophists were generally purveyors, not producers, of scientific
thought.[8] In eschewing specialization and abstruse research in the sciences and advocating instead a broad general knowledge, they were
bringing Greek educational practices into conformity with the age-old
national ideal and motto of pursuing "no interest to excess" (*mēden
agan*). It is well to bear these Sophists' innovations in Greek intellectual
life in mind because of the significance for the future that they were
destined to have.

Plato represents Socrates as having had an extreme dislike for the pretentiousness of Sophists in laying claim to knowledge in all fields. By contrast, Socrates pleaded ignorance as a basis of his inquiries and disclaimed any knowledge of the subjects professed by the Sophists. That Plato shared Socrates' distaste for the Sophists' exaggerated pretensions is evident from the way he describes their exhibitions of polymathy in his dialogues. On the other hand, he sided with the Sophists in disapproving overspecialization and in advocating broad training for the youth. He was also in complete agreement with Hippias' firm conviction that mathematical studies served as an excellent mental gymnastic. Plato's own enthusiasm for mathematics originated in his early studies with Pythagorean masters, Archytas among others. In Book VII of the *Republic* Plato discusses the Pythagorean quadrivium at length and regards it as an ideal propaedeutic in the education of philosophers.

The Academy that Plato founded reflected his intense zeal for mathematical studies. Over the portal were inscribed the words, "Let no one ignorant of geometry enter this building." His intrusion of mathematical discussions into his philosophical discourses must have upset many cultivated Athenians who attended his lectures. Aristotle reports an occasion when Plato drew a large crowd because he had announced that he was going to lecture on "The Good." When an audience expecting to hear his views on the good life or on wealth, good health, or some other favor of fortune found themselves listening to a lecture on mathematics, in which Plato concluded that The One was The Good, the enthusiasm of many of them for philosophy was dampened. Plato was able to pick his pupils and could turn away prospective philosophers who had no mathematical aptitude. In this way he succeeded in maintaining the key position of mathematical studies in the Academy during his lifetime; and at least one of his successors to the headship of the Academy is known to have been as firm as he in maintaining that policy.

Plato was himself not an original mathematician and is not considered an important figure among mathematicians, but because of the fame and popularity of his writings and the prominent position held by his Academy in the intellectual life of the ancient world, he has probably been more largely responsible than anyone else for the high favor that the study of mathematics has found in Western education through the ages.

Plato also made a unique contribution to the propagation of interest in the study of astronomy, mainly by his cryptic cosmological dialogue, the

Timaeus. This brief work which introduced Aegean Greek readers to Pythagorean concepts in mathematics, astronomy, and musical theory, and continued to intrigue Greek intellectuals down to the fall of the Academy in A.D. 529, left an incalculable heritage of doctrines and influences in the Middle Ages. Paul Shorey, staunch champion of Plato's importance in the intellectual development of the Western world, once remarked: "The shortest cut to the study of philosophy of the early Middle Ages is to commit the *Timaeus* to memory. Otherwise you can never be sure that any sentence that strikes your attention is not a latent quotation from the *Timaeus*, or a development of one of its suggestions."[9] Strong as this statement is, it is corroborated by the findings of the present study, drawn both from the writings that served as the antecedents of Roman science and from Latin science itself during the first Christian millennium. Because of its enigmatic style the *Timaeus* was much more widely known among the ancients through commentaries than through reading the original dialogue. As A. E. Taylor points out, *Timaeus* commentaries were published in Greece in every century from Crantor, who lived shortly after Plato, to Proclus, who, at the time of his death in A.D. 485, was head of the Academy.[10] As we trace the course of transmission of the traditions that molded Roman theoretical science, we shall see *Timaeus* doctrines coming to the fore at every turn.

Mathematical studies were always in a shaky position in the classical curriculum, and their cultivation required external stimuli. There never was any chance that Euclid's *Elements* would supplant Homer's poems as the cornerstone of Greek education or replace Virgil's *Aeneid* in the Latin schools. No sooner had mathematics gained an approved place in formal education than another teacher appeared who gave new attractiveness to literature and introduced a new group of subjects into the curriculum. Isocrates' efforts on behalf of his favorite studies proved successful, and accordingly he deserves to be ranked with Pythagoras, Hippias, and Plato among the founders of liberal arts traditions in Western education. Two factors aided Isocrates in his promotion of rhetorical studies: he could point to the obvious advantages of the study of rhetoric in the training of politicians, and he could show the intimate relationship between his subjects and the study that had always enjoyed greatest popularity among the Greeks, namely, literature.

Isocrates (436–338 B.C.) was a pupil of the eminent Sophists Prodicus,

Protagoras, and Gorgias, and received from them his enthusiasm for the study and teaching of rhetoric. After an early career as a writer of legal speeches, he took up the profession of rhetoric and taught oratory until his death fifty-five years later. He formalized the study of rhetoric and promulgated the idea that oratory should be regarded as a literary genre, and style cultivated as intensely in oratory as in other forms of literature. So successful was Isocrates in focusing attention upon style and elegance in oratory that the traditional emphasis in the Western world upon the cultivation of a literary style is attributable to him more than to anyone else. Perhaps because of Plato's influence, Isocrates relented in his opposition to the teaching of mathematics and came to acknowledge that geometry and astronomy performed useful services in keeping youth out of mischief while helping them acquire discipline and the ability to concentrate; but since mathematics had no practical applications, he disapproved of a prospective orator or philosopher continuing such studies into manhood.

Whereas Plato in the *Republic* had given mathematical subjects stress as preparing young men for careers as statesmen and philosophers, Isocrates was able to shift the emphasis to the rhetorical art. His influence was so potent that he set the keynote for education in the fourth century B.C. and—it would be no exaggeration to say—for all subsequent ages into modern times. At first the trivium of the rhetorical arts was introduced as the complement of the mathematical quadrivium, but it soon gained the dominant position; after but a brief period of prominence, interest in mathematical studies began to decline.

Shortly after Isocrates' death there emerged in Greece, as a result of the fame and leadership of eminent Alexandrian scholars, a great age of scholarship, chiefly of literary criticism and interpretation, succeeding the classical age of literary creativity. As literary genius declined and the heavy hand of pedantry began to cast a pall over Greek letters, the trend which originated with Isocrates, of inculcating in students canons of style and literary taste, strengthened; and formal education during the Hellenistic Age undertook to produce scholars as well as politicians and philosophers. To justify the place in the new curriculum of mathematical studies, which were losing their popularity in the schools, it was argued that a knowledge of mathematics would enable students to understand the recondite writings of the poets and philosophers.

By the end of the fourth century B.C. mathematical studies were still widely approved, alongside the rhetorical arts and literary training, as a part of the core curriculum of Greek youth aspiring to professional careers. Partly to meet the needs of secondary education and partly to gratify scholars who felt that a basic knowledge of science would assist them in comprehending the classical poets, a body of lay scientific literature began to appear, most of it compiled by scholars and schoolmasters who had little aptitude for scientific research but had developed some ability to make the work of others intelligible to students and laymen. Most original researches of Greek mathematicians and astronomers were too abstruse to be readily digested by these lay writers, but some salient features of the theories of leading scientists found their way in simplified—or perverted—form into the handbook tradition.

Good examples of the way in which some of the more digestible results of original scientific research might be incorporated in the lay handbooks while other more complicated theories might be rejected may be found in the case of Eudoxus of Cnidos (c. 408–355), who ranks with Archimedes as one of the most original and brilliant of Greek mathematicians. Eudoxus was a pupil of Archytas and a close associate of Plato, though probably not a member of his school. He invented and elaborated the general theory of proportions, later to be set forth in Books V and VI of Euclid's *Elements,* and he discovered the momentous Method of Exhaustion.

Eudoxus' claim to greatness as an astronomer rests upon a complicated system of twenty-seven concentric spheres which he devised to account for planetary motions and positions. It is regarded as the first truly scientific achievement in astronomy. The system was so ingenious and complex that it was not comprehended in modern times until Ideler's publications in 1828 and 1830 and not fully worked out in all its details until the celebrated Italian astronomer Schiaparelli published his study of it in 1875.[11] Naturally we would not expect to find any interest in this complicated system among popularizers and lay handbook authors. Perhaps this is the place to point out that Aristotle, a generation later, appropriated Eudoxus' theory and adapted and elaborated it into a mechanical system that reached the extravagant number of fifty-five concentric spheres—to account merely for the motions of seven planets and a celestial sphere. This system, despite Aristotle's reputation, suffered the understandable fate of a quick demise in antiquity. There were defects in it which made it unacceptable to ancient scientists and philosophers; but again we must not overlook the part that the handbook compilers had in its rejection. They found it, like Eudoxus' system, much too complicated to embrace. It is interesting that a Renaissance scientist, Fracastoro, made an abortive attempt to revive Eudoxus' system in his *Homocentrica*, published in Venice in 1538.

Some of Eudoxus' astronomical data did become established features of the handbook tradition. He estimated the period of a solar year as 365¼ days and of the retrogression of lunar nodes as 18½ years, and set the vernal point as the fifteenth degree of Aries. Because of a faulty observation he assumed a slight inclination of the sun's path to the ecliptic, and he estimated Mercury's maximum elongation at 23° and Venus' maximum at 46°. We cannot be certain that these figures, which became standard data in the popular handbooks, were actually introduced by Eudoxus, but the supposition is reasonable. In fact there is justification for ascribing to Eudoxus the same fundamental position in shaping the astronomical materials of the handbooks that the early Pythagoreans had in shaping popular mathematics.

Eudoxus' *Phaenomena*, the work that appears to have held this key position in popular astronomy, is lost; but we are informed, on the very good authority of Hipparchus, that the astronomical portion of the poem by Aratus bearing the same title was based wholly upon Eudoxus' work. Aratus' poem has already been referred to and it will be discussed at greater length later, because it presents the conventional features of the astronomy of lay writers and had an extraordinary vogue in antiquity. The handling of topics in the first two-thirds of the poem conforms to the treatment of the elements of astronomy in the popular handbooks. Aratus defines and locates the celestial circles—the parallels, ecliptic, zodiac, and Milky Way—identifies and locates the constellations, beginning with the Bears and taking up the northern constellations first, then those south of the ecliptic; and lastly records precisely which parts of the various zodiacal constellations will be observed to be setting on the western horizon as precise parts of other constellations are rising on the eastern horizon. He also records the ratio of the longest day to the shortest day for Greece.

Eudoxus wrote still another work, entitled *Periodos ges* ("Circuit of the Lands of the Earth"), which may have had a correspondingly important influence upon popular geographical writings, but the evidence here is tenuous. Existing fragments indicate that he approached the subject from the point of view of a mathematician. He also appears to have dealt with several topics of human geography—ethnology, religion, and medicine—that are not discussed in the more concise treatments of handbook geography. The very title of the work indicates the standard handbook procedure of using a coastal survey as the framework for dealing with geographical regions. Moreover, a number of doctrines widely circulated among ancient geographers were attributed by them to Eudoxus. It has frequently been assumed that Aristotle got his estimate of 400,000 stades for the earth's circumference (the earliest recorded estimate) from Eudoxus. Eudoxus is also reported to have estimated the length of the known world to be twice its breadth and to have surmised that the distance westward from Spain to India was not vast. The last two opinions became common property of the handbook literature. Eudoxus may be regarded as the forerunner of Eratosthenes, and it will be seen later that sufficient evidence exists to name Eratosthenes as the founder of the handbook traditions in geography. At the same time we must not ignore the possibility that the basic framework for handling geographical materials and many of the standard data of the handbooks may have originated with Eudoxus, who antedated Eratosthenes by a century and a half. Thus the indications are that handbook traditions were venerable and that basic patterns were laid down in an early period.

Greek science of the classical period comes to a close with the man who among the ancients enjoyed the highest reputation as a scientist. The towering figure of Aristotle (384–322 B.C.) must be reckoned with in the intellectual development of the Western world at every turn. He dominated the various fields of the natural sciences and shared with Plato the hegemony of philosophy until the Renaissance; but, paradoxically, he did not in antiquity occupy a comparably influential position in the theoretical sciences. He was neither professional mathematician nor astronomer, and mathematics is the science in which he appears to have been least competent. His curious mechanistic conception of a universe of fifty-five concentric corporeal spheres, set forth in Book XI of his *Metaphysica*, was a regressive step. But his adoption of the four Empedoclean elements and of the Pythagorean concepts of the perfection of spheres and circles, and his cogent arguments for a spherical, finite universe and for a spherical earth located at the exact center of the universe left an indelible impression upon the experts as well as upon lay scientific writers of antiquity. His doctrines about earth and water tending by nature to move toward the center of the earth, and air and fire tending to move away from the center, became common-places in popular scientific literature. Still other cosmological tenets extracted from his works are found widely scattered, like proverbial saws, throughout the handbook tradition. These distillations, sometimes mere traces, of the master's teachings found their way into Hellenistic handbooks not directly from Aristotle's works but through intermediaries, his successors in the Peripatetic school or other followers.

Aristotle's influence upon the development of the popular handbook movement was considerable; the presence of his doctrines will be noted as they crop up in later works. He had much to do with the increasing vogue enjoyed by technical manuals within a generation of his death. His pedantic approach to knowledge, the encyclopedic range of his interests, and his fondness for systematic treatment of subject matter—characteristics dear to handbook authors—were indelibly impressed upon his successors and pupils, who, as we shall see, figured prominently in the movement. Secondly, his bent for historical research, evidenced by his custom of introducing historical surveys of various fields into his own works as a basis for his investigations, led him and his pupils to study the doctrines of earlier philosophers. Nearly the entire body of pre-Socratic philosophy would have been lost to the ancient world had it not been for Aristotle. It is true that Plato had also been inter-ested in the pre-Socratics and had taught their doctrines in his school; some of our knowledge of pre-Socratic philosophy comes from Plato's works. But whereas Plato's interest was sporadic, Aristotle's was intense and continuing, and he made historical research a regular part of the program at his school. He engaged some of his pupils in compiling his-tories of various fields of knowledge: Theophrastus on physics and mathematics; Eudemus on arithmetic, geometry, and astronomy; Xenocrates on geometry, Meno on medicine, and Aristoxenus on the history of music. Except for fragmentary remains—extensive portions in the case of Meno's history [12]—these works have been lost; but they were so widely used as source material for handbooks that it has been possible to surmise the character and contents of some of them and, in the case of Theophrastus' history of physics, as we shall see, to make a substantial reconstruction of the work through a collation of later abstracts and quotations from it.

It would be hard to overestimate the significance of Aristotle's historical research and the historical works of his pupils, both for the handbook tradition and for the history of science itself. By transmitting to the Hellenistic world the doctrines of the Ionian physicists and the speculations of the early Pythagoreans, Aristotle and his pupils fired the imaginations of Hellenistic scientists and gave them new inspiration. An example of the influence of early speculations upon later scientific developments has already been pointed out—the share that the early Pythagorean notions had in the evolution of a heliocentric theory. Heliocentricism in a partial form persisted throughout the Middle Ages; and the very antiquity of Pythagorean doctrines seemed to give them a high degree of authority, even in the eyes of Copernicus. Some of the influences of the historical researches of Aristotle and his followers upon handbook science will be traced in the next chapter.

III

EARLY HELLENISTIC HANDBOOK TRADITIONS

Alexander's death in 323 B.C. brought about an immediate disintegration of his vast empire. A period of restive coalitions and internecine wars among his generals, provincial governors, and local princes resulted in a partitioning of his empire into separate monarchies. After dynasties had been established and internal affairs became somewhat stabilized, the jealousies among his successors continued to be manifested in their displays of majesty. The public buildings and temples at Alexandria, Antioch, and Pergamum served as façades for the Ptolemies, Seleucids, and Attalids. The enormous wealth of their domains was frequently diverted by vain rulers to increasing the splendor of those centers of culture. Reputable scholars and scientists became pawns in the rivalry, lured by stipends and court favors. They were also attracted by large collections of books in the national libraries and by excellent observatories and laboratories. Not until modern times was research again to experience such bountiful support from government funds.[1]

A new intellectual atmosphere was becoming discernible in Greece from the beginning of the Hellenistic Age: one of close study and reflection upon the works of the past—for literary experts an age of pedantic explication and of arranging the classical masters in canonical

lists to guide the tastes of readers; it was also a period of highly specialized scientific research. The proclivities of both scholars and scientists
provide sufficient resemblances to suggest that they were related phenomena of the age: the researches of both, conducted in ideal surroundings, were intensive, meticulous, and abstruse. The accomplishments of
scientists and scholars alike reflected glory upon their rulers and filled
the minds of lay contemporaries with admiration, when reports were
circulated. At the same time there is a striking difference to be noted
in the significance of their activities: whereas the philologists engaged in
heated disputes about the interpretation of the obscure meanings of the
poets and philosophers and argued the merits of writers of an earlier
age—occupations that tended to stifle literary creativity—the subsidies
poured into scientific research were producing a full flowering of Greek
scientific genius. Some results of the philological research became known
to the educated public through commentaries on the masterworks of
poetry, philosophy, and rhetoric; and some results of the scientific research through handbooks and commentaries of popular writers.

It is a curious phenomenon, and one that may have implications for
other times and places, that the Hellenistic Age (323–30 B.C.), which
is credited with producing the most significant scientific achievements of
the Greeks, is also to be debited with the full development of the popularization movement which, more than any other factor, was responsible
for the subsequent decay of ancient science. Intellectuals who derived
satisfaction from reading learned commentaries on Homer which treated
that poet as a sophisticated master of all knowledge also enjoyed perus-
ing popular manuals on scientific subjects which seemed to provide
insights into the great discoveries of Hipparchus and Archimedes. The
impression was being created that an understanding of latter-day scientific research would assist in the interpretation of early literary classics.
For these reasons it will be well to sketch here some of the more important accomplishments of Hellenistic scientists that filtered into the
lay handbooks.

Geographical knowledge was greatly increased during the early Hellenistic period. The organization required to administer Alexander's
enormous empire opened up new vistas in the east as far as the Ganges,
where Megasthenes (c. 290) was sojourning as ambassador of Seleucus
I. Ptolemy II reopened a canal connecting the head of the Gulf of Suez
with the Nile, and Alexandria became the port of transshipment for

cargoes from Spain and India. The coasts of East Africa and South Asia were becoming known to Greek mariners and geographers. Toward the end of the fourth century B.C. a Massilian Greek named Pytheas sailed around Spain, along the coast of Gaul, around Britain, and into the North Sea, perhaps as far as the Baltic—one of the more momentous voyages in the history of exploration. Of great significance to science was the fact that Pytheas was a navigator with a bent for astronomical observation; he kept careful records of the sun's declination in the northern latitudes of his voyage, which served as a basis for the fundamental work on mathematical geography written by Eratosthenes nearly a century later. After the results of his discoveries were adopted by scientific geographers, classic conceptions of a narrow belt of human habitation had to be discarded by lay writers. Henceforth astronomical observations were considered basic data for the cartography of large land areas.[2]

This period encompassed the careers of four of the most brilliant Greek mathematicians: Euclid, Archimedes, Apollonius of Perga, and Hipparchus—all of them intimately connected with research developments at Alexandria. It also witnessed the culmination of ancient heliocentricism in the truly heliocentric planetary system propounded by Aristarchus of Samos (c. 310–230). This theory marked the final step in the speculations of many men, directed, over a long period of time, at explaining the enigmatic motions of the planets. When the theory finally crystallized, even though it did away with the necessity of explaining planetary stations and retrogradations, it aroused little interest and soon withered away from neglect. Astronomy since the time of Plato had been predominantly the occupation of mathematicians, not of physical philosophers, and to a pure mathematician questions of physical reality were of little moment. In the next century the attitude and great reputation of Hipparchus doomed Aristarchus' heliocentric theory to virtual extinction in antiquity. Despite the precision of the instruments he was using, Hipparchus had been unable to find stellar parallax to corroborate an orbital motion for the earth. Besides, he was engrossed in mathematical procedures and calculations which would enable him to predict planetary positions that would coincide with his careful observations —an undertaking which fell short of its goal. For his investigations a geocentric orientation was the more convenient one.

Two brilliant medical researchers, Herophilus and Erasistratus, were laying the foundations for the scientific disciplines of anatomy and physiology at Alexandria early in the third century B.C. They performed and described minute dissections that were to provide a basis for the impressive anatomical studies of Galen four centuries later. So favorable were the conditions under which they worked—for the Ptolemies were generous supporters of medical research—that reports became current that the Ptolemies were turning condemned criminals over to them for vivisection. These reports have been doubted: they are reminiscent of stories of research assistance provided by Alexander to his former teacher Aristotle; and Galen's silence on the subject is hard to explain. More recently, however, Ludwig Edelstein has argued forcefully for their authenticity.[3]

Scientific investigations were but one manifestation of the general spirit of research and scholarship in this age. The originality and freshness that enhanced the creations of classical Greece gave way to laborious but unimaginative erudition. The very size of the collections of scrolls that had been accumulated in the libraries of Alexandria and Pergamum must have had a disturbing effect upon scholars studying there. Reputations were now to be made merely in coping with this accumulation of knowledge. The positions of librarian at Alexandria and Pergamum were reserved for men whose reputation for learning was pre-eminent. One poet of the time referred to the Alexandrian Museum as a bird cage in which scholars fattened up while engaging in trivial argumentation.[4] Orators and historians, epic, lyric, tragic, and comic poets were ranked according to current canons of taste—evaluations that were to have an important bearing upon which classical works would survive in modern times. Grammar and punctuation marks began to beguile the minds of intellectuals. The atmosphere of bookishness was so heavy and the desire to appear erudite in all fields so strong that several of the more gifted scientists took up the writing of compendia, along with their original researches, and by their reputations helped greatly to bring the handbook movement to the peak of its popularity.

Aristotle, who died at the beginning of the Hellenistic Age, may be considered the initiator of the intellectual spirit of the age. With his all-encompassing research interests and his pedagogical instincts, he set an example for scholars at the great centers of learning to investigate and systematize all fields of knowledge. Not long after his death the handbook movement was in full swing. His fondness for approaching studies from the historical point of view—a practice which he passed on to his followers—has an important bearing upon our knowledge of the history of ancient science. Much of the information that we possess was recorded by writers who followed Aristotle's precedent.

Theophrastus (c. 370–287), Aristotle's immediate successor as head of the Peripatetic school, served in that position for thirty-five years. One of his specialties was botany; in fact he is regarded as founder of the science of botany and he was the leading authority in the field until the sixteenth century. He was also a voluminous writer. Diogenes Laertius ascribed 227 treatises to him, on all sorts of topics. His works best represented Peripatetic doctrines to the Hellenistic Age. Theophrastus also became one of the great popularizers of science. His histories of arithmetic, geometry, and astronomy have been lost, but a more important work for the handbook tradition, his *Physical Opinions* in eighteen books, was so thoroughly ransacked by later compilers and epitomators that it has been possible to recover most of Books I and XVIII and to surmise the contents of the remainder of the work. This prodigious feat of scholarship was performed by Hermann Diels, who in 1879 published his reconstruction of this large body of received opinions that had been copied and digested in every generation in antiquity and are to be found everywhere in the handbook tradition.[5] Diels depended heavily upon a work by Stobaeus, *Eclogae physicae* (c. A.D. 470), which, when collated with an anonymous *Placita philosophorum* ("The Opinions of Philosophers," c. A.D. 150), revealed large-scale verbatim correspondence. Diels accordingly assumed a common source for both, named Aetius, whose book he supposed was derived from a compilation of the first century B.C., to which he gave the designation *Vetusta placita* ("Ancient Opinions"). This in turn he traced to Theophrastus' lost work on physics. Thus it appears that this bulky, so-called "doxographic" collection, originating in Theophrastus' work, was transmitted through the entire subsequent period of antiquity with little change. Posidonius, Varro, and Cicero are among the better known men who drew extensively on this collection of handy information.

Another pupil of Aristotle, Eudemus, at his master's instance wrote histories of astronomy and geometry which, together with Theophrastus' work on the subject, became the main sources of information on early developments in those fields. Eudemus' history of astronomy, for instance, is quoted by Dercyllides, who was a subsidiary source of a popular handbook by Theon of Smyrna (c. A.D. 125), which will be discussed later. The fullest information about Eudoxus' concentric system comes from a commentary on Aristotle's *De caelo* which Simplicius wrote near the middle of the sixth century. Simplicius quotes largely from Sosigenes, a second-century Peripatetic, and remarks that Sosigenes drew his information from the second book of Eudemus' history of astronomy. Eudemus' companion work, the history of geometry, held a similarly important position as a source of information about the evolution of geometry up to Euclid. The so-called "Eudemian" summary of the early history of geometry, found in Book I of Proclus' *Commentary on Euclid,* had been commonly assumed to be an extract from Eudemus' work, but Heath pointed out that though Eudemus was the source he was not the original author of the summary.[6]

It was suggested earlier that Eudoxus' scheme of twenty-seven concentric spheres and Aristotle's scheme of fifty-five spheres were too complicated to find favor with popular writers. In addition they both had another defect that tended to make them unacceptable. In placing the planets on concentric spheres whose center was the earth, Eudoxus and Aristotle did not allow for variations in the planets' brightness resulting from variations in their distances from the earth. A habitual observer of the night sky soon becomes aware of the great range of difference in luminosity as the more conspicuous planets move from apogee to perigee. Venus, for instance, is $2\frac{1}{2}$ times brighter at positions near perigee than at positions near apogee, her apogees sometimes being 160 million miles and perigees only 26 million miles. The credit for accounting for this phenomenon goes to Heraclides of Pontus (c. 385–315 B.C.), who like Eudoxus was a pupil of Plato. Heraclides was more of a polymath than a scientist and is not to be classed with Eudoxus as a mathematician and astronomer; yet his scheme explained the behavior of Venus and Mercury so neatly that it won immediate approval and cast disfavor upon concentric systems. He suggested that the revolutions of Venus and Mercury were not about the earth but about the sun, and that the sun and the other planets were orbiting around the earth. This geoheliocentric theory represented a giant step toward the complete heliocentric theory which was developed a century later by Aristarchus of Samos. Heraclides also maintained another assumption which would become implicit in the heliocentric hypothesis: that the great diurnal rotation of the heavens was apparent and not real, and was to be accounted for by the earth's diurnal rotation on its axis.

These ingenious and startling hypotheses propounded by Heraclides and Aristarchus served only to confuse and confound astronomers on all levels. A partial or complete heliocentric theory accounts very nicely for variations in the brightness of planets and for other enigmatic phenomena—such as stations, at regulated points of maximum elongation, and retrogradations—which seemed to indicate that the sun was controlling planetary motions. But doctrines of separate planetary spheres at fixed distances from the earth had been so firmly embedded in Pythagorean, Platonist, and Aristotelian concepts and in the popular literature that it was hard to prevail against them. Writers on astronomy who should have known better are found introducing irreconcilable geoheliocentric views into systems of fixed planetary spheres. Authors of popular handbooks generally embraced Heraclides' view of the heliocentric motions of Venus and Mercury; but they rejected both his idea of the earth's diurnal rotation on its axis and Aristarchus' heliocentric theory as incredible, or at least inadmissible into an elementary system designed for general readers. It is interesting that, although Aristarchus' remarkable system received almost no notice from popular writers, his estimate of the sun's distance from the earth as between 18 and 20 times the moon's distance, in his small treatise *On the Sizes and Distances of the Sun and Moon*, did receive wide notice in the handbooks. As so often happened, the minor treatise has survived while the works containing the remarkable speculations have perished.

Like Eudoxus, the celebrated geometer Euclid (c. 330–275) also wrote a textbook with the title *Phaenomena*, a collection of eighteen theorems and proofs, which was to exert a powerful influence upon compilers of popular handbooks on astronomy. It belongs to a corpus of textbook writings on astronomical spherics which is thought by historians of mathematics to have first taken shape in the middle of the fourth century B.C. and which may have been related to Eudoxus' lost *Phaenomena*. There are marked resemblances in treatment and in geometric propositions between Euclid's textbook and the treatises *On the Moving Sphere* and *On Risings and Settings* by Autolycus of Pitane (c. 310 B.C.), whose works form a part of the corpus and are the earliest complete mathematical works that have survived from Greek antiquity. The entire corpus of theorems and proofs is confined to the fixed stars. It deals with their risings and settings and daily motions, the celestial great circles—parallels, horizon, meridian, zodiac, and Milky Way—and proofs that the universe is spherical and the earth is at its center. Treatises on astronomical spherics, such as those by Eudoxus, Autolycus, and Euclid, appear to have exerted a strong influence upon popular astronomy: many classical writers devote greater attention to the celestial sphere and its circles and the location of the earth than they do to planetary motions. But although the works of the early mathematicians were presented in the form of geometric propositions and proofs, the treatments of popular writers merely appropriated the results.

One cannot leave Euclid without mentioning another textbook of his, the celebrated *Elements*. One of the masterworks of scientific literature, this book is to be classed with ordinary handbooks in one respect alone, namely that a large part of its materials was drawn from earlier sources. In all other respects, and in particular in its superbly logical arrangement of theorems and development of proofs, it displays the genius of its author. It also contains some original materials. George Sarton calls the *Elements* "a monument which is as marvelous in its symmetry, inner beauty and clearness as the Parthenon, but incomparably more complex and durable." [7] It is one of the finest and certainly the most successful textbook ever to have been produced. At a time when interest in mathematical studies was beginning to wane, Euclid's volume was a powerful force in retarding the trend. It enjoyed an immediate and continuing vogue. Perhaps no work in the Western world except the Bible has been so widely studied and edited. It has been the basis of geometry studies through the ages. English schoolboys in the present century were still using a translation of the *Elements* that did not differ greatly from the original work. However, Euclid's extraordinary talent for making a difficult subject intelligible did not discourage later mathematicians—such as Heron, Pappus, Porphyry, Proclus, Simplicius, and several Arabs—from publishing commentaries on the book.

The reception accorded this purely mathematical work in an age when mathematical studies were beginning to be related to literary exegesis was not wholly to be expected. A more typical product of the age was the astronomical poem already referred to, the *Phaenomena* of Aratus of Soli (c. 315–240). The author is a prime example of a littérateur dabbling in science. He was invited to sojourn at the courts of two Hellenistic monarchs, and delighted both of them with his poetry and learning. Though he had no competence in astronomical studies, Aratus undertook a poetic version of Eudoxus' *Phaenomena* which, coupled with a poem on weather signs drawn mainly from Theophrastus' treatise on the same subject, succeeded in making the poet the most celebrated authority on astronomy in antiquity. As Euclid was the geometer, Aratus came to be regarded as the astronomer—a rather ridiculous identification in the latter case but not hard to understand when we consider the confusions arising from the literary affectations of some scientists and the scientific pretensions of many littérateurs of that time. Because of its popularity Aratus' literary curiosity has survived, whereas nearly all the great works on astronomy from antiquity have been lost.

Aratus' poem is divided into three nearly equal parts. It is easy to see why the first two parts, known as the *Phaenomena*, enjoyed great acclaim. The treatment of the constellations never becomes technical. The phraseology is picturesque and figurative, and Aratus makes frequent use of the rich store of epithets dear to classical poets and mythographers. At a time when mathematics and theoretical science were receiving less attention in the schools and, correspondingly, pretensions to scientific knowledge were becoming more pronounced, Aratus gave readers the exquisite pleasure of seeming to be grasping a scientific discipline in a delectable poem. The third section of the poem was distinct from the rest and dealt with weather signs; separately it was referred to as the *Prognostica*. Inspired by Hesiod's poem on farming, it too attracted many readers and admirers with its helpful hints to farmers. Virgil used it extensively in his *Georgics*. Aratus had considerable influence upon Lucretius and other Latin poets. Ovid predicts, in a verse in the *Amores* (I.15.16), that Aratus' fame will last as long as the sun and moon. St. Paul in Acts 17:28 was quoting a verse from Aratus' invocation.

Another indication of the popularity of the *Phaenomena* is the fact that four Latin translations of it, of small intrinsic merit, have survived. In addition to fragments of a translation by Varro Atacinus, we have 769 lines of Cicero's translation, prepared when he was a young man; 857 lines of a translation by Germanicus Caesar, a nephew of the emperor Tiberius; and 1,878 lines of a paraphrase by Avienus, a didactic versifier who lived in the fourth century.

Aratus' poem was intended to be an easy guide to the heavens for readers who had no technical instruction. Its enormous popularity, rather than the complexities of its subject matter, accounts for the fact that a train of commentators saw fit to enlarge and expound upon the poem. The names of twenty-seven commentators on Aratus are known to us, and four of the commentaries have survived. Euclid's *Elements* also inspired a host of commentators; and the extant commentaries on Euclid and Aratus add a great deal to our knowledge of early mathematics and astronomy. But it is sad to reflect upon the fortunes of most of the ancient scientists when contrasted with those of the popularizers. As has already been pointed out, the one work of the great astronomer Hipparchus to survive is his commentary on Aratus' poem. Had it not been for Aratus' popularity, we would have had nothing from the pen of Hipparchus.[8]

Eratosthenes of Cyrene (c. 275–194), the next to be considered here, has been an enigma to historians of science.[9] He clearly deserves ranking with the great scientists of antiquity, yet in his own day his reputation was based mainly upon the prodigious range and expertness of his erudition. His career was in line with the fashions of the time. Reputations could be made much more readily in learning than in science, especially at Alexandria; and Eratosthenes, a naturally versatile person, found it easy to be fashionable. His repute as a man of learning rather than his reputation as a scientist was responsible for the frequency and durability of his doctrines in the popular handbooks on science. One bit of cogent evidence for this statement may suffice; although Eratosthenes' geographical data and views were roundly and effectively criticized by Hipparchus, they were much more influential in later ages than those of Hipparchus.

Eratosthenes possessed the ideal components—conspicuous accomplishments in various fields of science and an overpowering reputation in learning and scholarship—to make him the gigantic figure that he became and remained in the traditions of lay handbook science. In mathematics he proposed a solution for the famous Delian problem (duplication of the cube), devised the "sieve of Eratosthenes" for finding successive prime numbers; wrote a work *On Means*, in two books, which was highly regarded by Pappus; and he had the signal honor of having an important work by Archimedes, an older contemporary, dedicated to him. Eratosthenes' most noteworthy scientific accomplishments were in the field of geodesy. He devised a method of measuring the earth's circumference by marking the angle of the sun's declination at two widely separated points on the same meridian and using geometrical methods to determine the number of degrees of arc between the points. His final result, a figure of 252,000 stades (less than fifty miles from actuality, according to a value of the stade assigned by some scholars [10]), was, to be sure, a fortuitous calculation, because it involved some crude measurements and geodetic observations and was intended to be only approximate. Other geodetic researches and his subsequent mapping of the known world on a framework of supposedly scientific latitudes and meridians established his merit as the founder of scientific geography and made him the most celebrated figure in that field until Ptolemy. One of his works on mathematical geography also dealt with the distances and sizes of the sun and moon. This investigation naturally attracted a great deal of attention and admiration in the works of lay authors and his figures got into the doxographic collection. As we shall see later, the handbook traditions of Western Europe almost completely bypassed Ptolemy; consequently, Eratosthenes continued to be considered pre-eminent as a mathematical geographer by the few medieval men in Western Europe who interested themselves in such matters. He was a venerable authority, and little more, to Latin scientific writers in the early Middle Ages, many of whom did not know how to spell his name but could give his figure for the earth's circumference correctly.

Eratosthenes was acknowledged to be the most learned man of his day and he undoubtedly was one of the most learned men of antiquity. For the last forty years of his life he served as librarian of the Alexandrian Library—the top intellectual position of the Hellenistic world—and he acquired a reputation for versatility that was proverbial. His detractors dubbed him "Beta" or "Pentathlos," suggesting that he attempted too many specialties and was of first rank in none. What endeared him to the textbook writers and compilers was the fact that he was primarily a man of letters and often used his acumen in the sciences to explain obscure matters in literature and philosophy.

He was an influential figure in many fields of handbook science. His work on mathematical concepts in Plato's philosophy, entitled *Platonicus*, has been lost, but enough is known about it to indicate that it was essentially a commentary on the *Timaeus* and in content and form resembled handbooks in this field that have survived. It dealt with arithmetical proportions and the different kinds of means, together with the numerical ratios in concordant sounds, as a basis for comprehending Plato's discussion of the fabrication of the World-Soul and his doctrine of the harmony of the spheres. It may also have included a section on cosmography. It was a leading source of Theon's handbook and is cited there twice by author and title.

It is in mathematical geography that Eratosthenes' dominant position in handbook science is most clearly seen. The evidence indicates that he molded the traditions of that subject, possibly taking much of the arrangement and content of his geography from Eudoxus. Eratosthenes' *Geographica* has been lost, but a detailed discussion of the contents of its three books may be found in the opening books of Strabo's *Geography*. Book I of Eratosthenes' work was a historical account of geographical conceptions, going back to Homer and Hesiod. Book II was devoted to mathematical geography, as we learn from various sources besides Strabo, who had an obvious dislike for the subject yet ventured to criticize Eratosthenes. A listing of the more important features of Eratosthenes' geodetic and cosmographic materials will serve to persuade readers who are familiar with the works of ancient geographers that it was Eratosthenes who laid down the principles.

Book II opened with a discussion of the arguments for a spherical earth and universe and gave the location of the celestial circles and the earth's inhabited quarters in the upper and lower hemispheres. Eratosthenes set latitudes and meridians for the known world, locating his latitudes at the Cinnamon Coast (northern extremity of Somaliland), Meroë, Syene (Aswan), Alexandria, Rhodes, the Troad, the mouth of the Borysthenes (Dnieper), and Ultima Thule; he estimated the distance of each latitude from the equator, and recorded the number of hours of daylight on the longest day of the year at each latitude. His distances were based upon his figure of 252,000 stades for the earth's circumference. The book included an account of the methods he used to ascertain this measurement. Eratosthenes also estimated the length of the known world from east to west, assuming that the habitable lands known to the Greeks were surrounded by a continuous body of ocean. Although he was generally critical of Homer's geographical notions, he may have been influenced by the durable Homeric conception of an ocean encircling the world. Herodotus had maintained that there was no evidence for a circumambient northern ocean. Strabo says that Eratosthenes was persuaded there was one by reports of tidal phenomena observable in both the Western Ocean and the Red Sea. At any rate it was Eratosthenes' notion of a circumambient ocean that prevailed among popular writers. He also argued that the size of the highest mountains was negligible in comparison with the size of the earth.

These basic conceptions of Eratosthenes and their underlying proofs became stock features of the handbooks, which were transmitted from generation to generation with little change. Distant expeditions, and soon the Roman conquests, were to add a substantial amount of new data and radically alter ancient geographical ideas. Yet most handbook authors preferred to retain Eratosthenes' conceptions and figures, introducing such minor revisions as shifting the southern extremity of human habitation closer to the equator, to reflect the discoveries of expeditions conducted on the Upper Nile, or offering an alternate estimate for the earth's circumference, one proposed in the first century B.C. by Posidonius. It gave readers a comfortable feeling, in their wonder at the achievement, to be able to choose between two figures, even though the estimate of Posidonius was nearly a third less than that of Eratosthenes.

The last noteworthy figure of the early Hellenistic period to be discussed here serves as a fitting climax to an age in which polymaths and literary critics appropriated the findings of creative scientists and plausibly represented them to the world at large in handbooks and commentaries. Crates of Mallos (fl. 160 B.C.), first librarian of the famous library at Pergamum, founder of the Pergamene school of criticism, and author of the first formal treatise on grammar, got his reputation from the heated debates he carried on with eminent Alexandrian critics about the proper interpretation of Homer's poems. Since the time of Chrysippus, successor (232 B.C.) of Cleanthes as head of the Stoic school of philosophy, that school had become interested in the interpretation and expounding of Homer; and Crates, a leading representative of the school, came to regard Homer not as an inspired poet but as an omniscient and obscure master of the secrets of the physical world, secrets veiled in allegory but explicable through the erudition of a commentator. Virgil's *Aeneid* underwent similar treatment in Western Europe during the Middle Ages. Crates' zeal to expound Homer's meaning in the light of Stoic physical theories involved him in considerable writing on scientific subjects, although he had no training as a scientist. His symmetrical disposition of the earth's great land masses continued to be one of the dominant conceptions of *mappa mundi* cartography through the Middle Ages and even exerted an influence upon geography in the early Renaissance.

Crates conceived of the earth as a globe with four inhabited quarters, transversely and diagonally opposed to each other; these were separated from one another by two belts of ocean, one going longitudinally about the poles and dividing the eastern and western hemispheres, the other about the equator, dividing the northern and southern hemispheres. The human habitations were assumed to be quite alike, with corresponding conditions for maintaining life. The collision of ocean's streams at the poles and the resulting backwash caused the ebb and flow of the tides. This likely conception of the earth (ostensibly based upon Homer!) had much to recommend it to the fancies of later readers. It was given currency by Posidonius and Cicero; through Macrobius and Martianus Capella in the fifth century it was transmitted to the Middle Ages until, by the twelfth and thirteenth centuries, it had become one of the most popular conceptions of the earth to be held by cosmographical writers. Still later, Crates' notions also played a prominent part in stimulating interest in the exploration of lands beyond the oceans. Thus we find a Hellenistic literary critic who had attributed his conception of the earth's habitations to Homer influencing cartography with it for a millennium and a half.

By the middle of the second century B.C. scholars, critics, and grammarians had taken over education and the dissemination of scientific knowledge. Political strife and anarchic conditions in Egypt had had a detrimental effect upon scientific research and scholarship at the Alexandrian Museum and Library. The persecutions of Ptolemy VIII, who came to the throne in 146, drove many Greeks from Alexandria. According to one ancient writer [11] the eastern Mediterranean area became filled with refugee grammarians, philosophers, geometricians, musicians, painters, and physicians. Many of them were impelled by hunger to take up the poorly paid profession of teaching; thus the textbooks of their disciplines gained still wider circulation, the effects being felt as far as Rome. For several centuries Alexandria no longer enjoyed the primacy it had held over leading eastern centers of learning at Antioch, Smyrna, Pergamum, and Rhodes.

IV

THE POSIDONIAN AGE

The Greco-Roman age of civilization began in the second century B.C.
During that century contacts between intellectual leaders of both nations became so impressionable that the cultures of the eastern and
western Mediterranean areas were never again to be the same. For this
welding together of two cultures that were sharply different the Romans
were more directly responsible than the Greeks. Immediately after the
defeat of Carthage in 202 B.C., the Roman Senate embarked upon a
policy of intervening in Greek affairs. Ever since Alexander's death the
states formed from the partition of his empire had been suffering the
effects of bitter jealousies and strifes. Roman intervention was now frequently welcomed by rulers who were so eager to destroy their rivals
that they were willing to risk the consequences of soliciting Roman
military help. When Roman legions were dispatched to participate in
internecine conflicts, they usually conquered the Greek forces with surprising ease. Several Greek states sent deputations, some of them staffed
by philosophers, to petition the Roman Senate for aid or, if they had
already been conquered, to seek more favorable truce terms. In a little
more than fifty years Roman armies had subdued all of Greece.

Sober-minded Greeks could not help being deeply impressed by the marked military superiority of Roman legionaries, fighting on behalf of an ostensibly representative form of government, over Greek soldiers who were supposedly democratic by instinct and lovers of freedom but had become despondent and spiritless under the long domination of ambitious despots. Naturally there was a widespread inclination to admire both the Roman republican form of government and its political leaders.

During this period a number of Greek intellectuals had occasion to visit or live in Rome. The librarian Crates was dispatched to Rome as an envoy of King Eumenes II of Pergamum (c. 168 B.C.). The historian Polybius was one of a thousand Achaean citizens who were deported to Rome as hostages after Aemilius Paullus' victory over King Perseus of Macedon. Polybius did not suffer imprisonment with his fellows but was invited to the home of Paullus to become the tutor of his sons. Paullus arranged for the removal of Perseus' library to his home in Rome in 167; thus, for the first time, Romans interested in Hellenic culture had access to a sizable collection of Greek writings. Scipio Africanus the Younger, son of Paullus, organized a literary and philosophical circle to stimulate Roman intellectuals to a study of Greek literature and explore ways of introducing Greek studies and adapting them to Roman cultural pursuits.

Polybius and the Stoic philosopher Panaetius were invited to join the Scipionic Circle and for a number of years exercised considerable influence upon the other members. A warm admirer of Scipio, Polybius wrote an elaborate history to explain how Rome had brought the civilized world under its domination. He had been eyewitness of momentous Roman conquests, having been invited by Scipio to be present at the destruction of Carthage (146) and Numantia (134). His explorations along the North African coast and into Gaul and Spain provided him with the insights which enabled him to become the first universal historian. His example of stressing geographical knowledge as an important factor in historiography left an indelible impression upon later historians and geographers. Panaetius became the leading Stoic philosopher of his day and the most potent force in developing in the Stoa the syncretistic tendencies which were to culminate in Posidonius a century later. The development of his philosophical outlook was strongly influenced by his associations with Roman noblemen, whom he admiringly regarded as applying philosophical ideals in their careers of exemplary and effective action. In 155, when Athens found herself in disfavor with the Roman Senate, she sent as special envoys the heads of her three leading philosophical schools—Critolaus the Peripatetic, Diogenes the Stoic, and Carneades, founder of the New Academy. During their stay these men delivered public lectures which did much to stimulate Roman intellectuals to the study of philosophy.

Thus began a lengthy period of mutual admiration—Greek intellectuals basking in the late-afternoon sunshine of Roman adulation, being feted at the salons of Roman nobility and political leaders, inculcating in the minds of their Roman hosts an enthusiasm for Greek philosophy and literature, while themselves conceiving admiration for the vigor of these men of affairs and seeking new ways to ingratiate themselves with these appreciative dilettantes. Sojourns in Rome had much to do with turning the minds of Greek philosophers from the passive to the active virtues and with molding eclectic philosophies which would have enough application and adaptability to appeal to practical men in this dynamic new world that was coming into being.

The Stoic school had a prominent part in the trend of philosophies to enlarge their scope and to become all things to all men. Panaetius and his pupil Posidonius readily embraced Platonic and Aristotelian tenets that seemed more adequate than early Stoic attitudes to explain the convulsive changes that were happening. Stoics began to see in the new world empire of the Romans a fulfillment of the commonwealth of God, in which the citizens were cosmopolites and distinctions between East and West, Greek and barbarian, slave and free, philosopher and layman were passing away.

These tendencies reached their culmination in Posidonius (c. 135–51 B.C.), the foremost Stoic of his day. There have been several notable efforts in the past generation to reduce the estimates of Posidonius' influence upon Roman philosophy and literature,[1] but in his relation to the lay handbook tradition his prominence must be reaffirmed. His key position in the intellectual developments of the Greco-Roman world has been compared with Aristotle's prominence in an earlier period; and nothing less than that sort of comparison would be adequate to indicate the influence of Posidonius upon later compilers. Aristotle gathered, systematized, and embodied the knowledge of his day in a large collection of writings which, together with the treatises of his pupils, transmitted this vast body of knowledge to the Hellenistic world.

Posidonius was the great popularizer of learning for the Greco-Roman world, and because of his encyclopedic interests and his ability to present information in a readily digestible form, he became the chief transmitter of Greek philosophy and science to the Romans.

The vigorous activity that marked Posidonius' early life afforded him a variety of experiences that were to determine the course of his career. A native of Syria, he spent a number of years at Athens, absorbing the modified Stoic doctrines of Panaetius. After his teacher's death he embarked upon an extended tour in which combined geographic, ethnographic, and historical studies of the places and peoples he visited along northern Mediterranean shores as far as Cadiz and along the North African coast into Egypt. He settled at Rhodes and established a school of his own, which soon became renowned as a center of Stoic learning. Many wealthy young Romans who had come to Greece to complete their academic training attended it. In 86 Posidonius was sent on the first of his diplomatic missions to Rome, to plead on behalf of the Rhodians before Marius. In 78 Cicero attended his lectures at Rhodes; and Pompey visited him twice, in 67 and 62, to hear him lecture and to pay him homage.

No work from Posidonius' voluminous collection has survived, but the form and substance of some of his writings can be partially surmised from digests and extracts made by later authors. His book on meteorology (*Peri meteoron*) was used extensively by Lucretius in his poem *On the Nature of the Universe*; and inasmuch as it was also the main source of Seneca's *Natural Questions* and of the scientific parts of a poem *Aetna*, formerly ascribed to Virgil, we have some notion of its contents. Similarly, the scope and matter of another work, entitled *Regarding the Ocean* but dealing mainly with geography, may be conjectured from the many notices of it in Strabo's *Geography*. Other works are represented by scanty fragments and some only by title. Posidonius' *Histories*, in fifty-two books, was probably his most important work. A continuation of Polybius' *Universal History*, it was, like its predecessor, broad in scope. It encompassed events in the East and West from the destruction of Carthage to the dictatorship of Sulla. The *Histories* became for the Romans a model of historiography, introducing them to concepts of universal history and demonstrating the importance of geographical and ethnographic considerations in the interpretation of history. Posidonius' ethnographic studies of Gallic and North African peoples are thought to have helped develop the ethnographic interests shown by Caesar in his *Gallic War* and by Sallust in his *Jugurthine War*. Like Polybius, Posidonius viewed Rome's establishment of a world empire as fulfillment of the providence of God, a divine Stoic commonwealth on earth. He showed strong partiality toward Pompey and the Roman nobility in his *Histories* and added a lengthy appendix on Pompey's eastern campaigns which supplied Roman writers on geography with data about the Levant. Posidonius exercised a particularly potent influence upon Roman philosophical literature. His intimate friend and admirer, Cicero, used his works extensively in composing *On the Nature of the Gods* (Book II), *On Divination* (Book I), *On Duty* (Book III), the *Tusculan Disputations* (Book I), and the brief *Dream of Scipio*. It is mainly owing to Posidonius that later Roman Stoics like Seneca and Marcus Aurelius display such strong Platonist tendencies. In fact, Cicero's eclecticism is but a further extension of that which had been rife in Greece for two centuries and had come to the fore in Posidonius.

Some appreciation of the preponderant influence wielded by Posidonius upon Roman writers who handled scientific subjects may be gained by comparing Greek works that are assumed to have been largely derived from Posidonius with the Roman handbook literature. A considerable amount of doctrinal information and data, extracted or digested from Posidonius' works by such Greek writers as Geminus, Cleomedes, and Theon of Smyrna, corresponds or bears striking similarity to materials in the Roman handbooks. Instances will be given when the writers are discussed separately.

Posidonius' *Peri meteoron*—the main source of Seneca's *Natural Questions*, as has been mentioned—was based upon Aristotle's *Meteorologica*. Geminus wrote a commentary on the *Peri meteoron*, and some of the many citations of Posidonius by Cleomedes may also have been drawn from it. From Strabo and others we gather that Posidonius' *Regarding the Ocean* dealt with mathematical as well as ordinary geography, gave estimates of the breadth of zones, discussed human habitations, following Crates in the assumption of the existence of peoples diametrically opposite Europe in the southern hemisphere, and also assumed that a continuous body of water surrounded the known inhabited world. Posidonius' commentary on Plato's *Timaeus* was probably the most influential in the long line of commentaries on this work, serving as a main source of the *Timaeus* commentaries of Plutarch, Theon, Porphyry, and Chalcidius, and being largely responsible for the revival of the Pythagorean arithmetic and number lore which figure prominently in the works of many Romans writing on a variety of subjects. This revival, as we shall see, was to lead to the publication of Nicomachus' work on arithmetic (c. A.D. 100), and thereafter arithmetic rivaled and sometimes superseded geometry as the topic of discussion in the mathematical sections of popular handbooks.

The fact that none of Posidonius' works has survived adds to the difficulties of trying to assess his true worth as a man of science, a matter on which the opinions of historians differ greatly. Apparently he was a competent mathematician. Several definitions in elementary geometry are attributed to him by Proclus, who also reports that he distinguished seven species of quadrilaterals and expressed himself on the distinction between theorem and problem. Again Proclus tells us that Posidonius wrote a book in refutation of a work by an Epicurean named Zeno who had objected to some of the basic principles of Euclid's *Elements.*

Other more conspicuous and influential accomplishments in science are credited to Posidonius. Long before his time various writers had drawn attention to a possible connection between tidal phenomena in eastern and western waters. Nearchus, Alexander's admiral, had reported tides in the Indian Ocean; and information about tidal phenomena in the Red Sea and Persian Gulf had been circulated among early Hellenistic navigators. Pytheas had noted an influence of the moon upon tides during his voyage in the Atlantic at the end of the fourth century B.C. Seleucus, of Seleucia on the Persian Gulf—who, though called a Babylonian by Strabo, was more probably a Greek—made observations of the tides in the Red Sea and connected them with positions of the moon. He offered a quaint explanation: that the moon's motion deflected the atmosphere and set up a resistance to the rotation of the earth's atmosphere. It remained for Posidonius to be the first man, so far as surviving evidence indicates, to reach conclusions about tidal phenomena that conform with modern views. During his thirty-day stay at Cadiz, on his early peregrinations about the Mediterranean, he kept careful records of ocean tides. He observed that both the sun and moon are factors in producing the monthly variations. The tides are fullest at new moon, he noted, when sun and moon are in conjunction, and at full moon, when they are at opposition. He attributed these phenomena to terrestrial and celestial sympathy. His observations, coupled with his knowledge of tidal phenomena in eastern seas, led him to conclude that one continuous ocean surrounded the known world. The weight of Posidonius' authority, added to that of Eratosthenes, made it certain that the notion of a circumambient ocean would be a dominant concept in world geography among Latin geographers. Besides, Homer's concept of "backward-flowing Ocean" could not be easily discarded. In the eyes of popular writers Homer was still a master of all knowledge.

Another scientific investigation of Posidonius aroused widespread interest, and his great authority caused his findings to be circulated in the handbooks and to be adopted even by eminent scientists. He undertook a new measurement of the earth's circumference. His calculations were based upon an observation of the meridian altitude at Alexandria of the star Canopus which he observed to culminate on the edge of the horizon at Rhodes, and upon a rough estimate of the distance between the two places. His first result was a figure of 240,000 stades, not very different from Eratosthenes' figure of 252,000; but he revised his estimate to 180,000 stades. It has been conjectured that his original estimate was based upon a figure of 5,000 stades as the distance between Alexandria and Rhodes and that he later used 3,750 stades for the distance; or that he was using two different values for the stade. D. R. Dicks rejects these and other attempts to apologize for Posidonius' grossly inaccurate later estimate.[2] An account of the methods of both Eratosthenes and Posidonius is to be found in Cleomedes' handbook.

The great discrepancy between the figures of two highly respected authorities posed a problem for popular writers, many of whom avoided the difficulty by reporting both figures and letting the reader choose. Strabo, Varro and Pliny preferred Eratosthenes' estimate, a figure that had become well established in the handbook traditions; and because Varro and Pliny were highly regarded as authorities by Latin scientists, Eratosthenes' figure was widely adopted in the West during the Middle Ages. On the other hand, some indication of the weight of Posidonius' authority among scientists is seen in the fact that Ptolemy adoped Posidonius' over-all figure as a basis for his mensurations of the known world in his classic work on geography. Ptolemy's diminished conception of the size of the world, in turn, had an important bearing upon Columbus' decision to attempt a westward sailing. But whenever we find the estimate of 180,000 stades adopted by writers in the Latin West in the early Middle Ages, it is to be assumed that Posidonius and not Ptolemy was the authority, for Ptolemy had virtually no influence upon Latin writers until Arabic and Greek works began to be introduced in the West early in the second Christian millennium.

Still another bold attempt at ascertaining dimensions—this time of the sun—enhanced Posidonius' reputation as a scientist. According to Cleomedes, he assumed the sun's apparent orbit to be 10,000 times as great as the earth's circumference. This figure he had obtained from an estimate used by Archimedes in his *Sandreckoner*. Actually Archimedes, at that time, was attempting to prove that the sun's orbit must be *less than* 10,000 times the earth's circumference. Accepting a report of Eratosthenes that at Syene, beneath the Tropic of Cancer, the sun, at the summer solstice, casts no shadow over an area 300 stades in diameter, Posidonius concluded that the sun's diameter would be 10,000 times as great as the diameter of the shadowless area. His methods and assumptions are naive and crude, as were all attempts to measure the sun before the age of scientific instruments; but his result—an estimate of the sun's diameter as 39¼ times the earth's diameter—happens to be a fairly good one by ancient standards, much better than the estimates of Aristarchus, Hipparchus, and Ptolemy. Posidonius' procedures did not receive much notice in the Latin works on science, but faint traces of them appear occasionally.

Despite his conspicuous record of accomplishments, historians of science have been chary of giving prominence to Posidonius as a scientist because the total effect of his career upon subsequent developments was a bad one.[3] To be sure, his influence upon geographers, in stimulating interest in human geography, and upon historians, in revealing the importance of considering ethnography and geography in interpreting historical events, was wholly salutary. Like Eratosthenes, Posidonius was reputed to be the most learned man of his day, and Strabo and Galen later attested to his top ranking in the intellectual world. Living in a still more dilettante age than Eratosthenes, he bore many resemblances to that polymath. But in a much larger measure than his predecessors Posidonius had the curse of the popularizer upon him. His impressive compendia, readily digestible and chock-full of plausible information and scientific explanations, led to the neglect of original research. After Posidonius, the hand of written authority became very heavy, in the West too heavy to be lifted again for over a thousand years. He had a large part in determining what views were to survive in later ages. We have noted that his estimate of the earth's circumference prevailed in scientific circles to the time of Columbus, through Ptolemy's adoption of it. Seleucus, Posidonius' predecessor in the lunar explanation of tidal phenomena, was also one of the few authorities who accepted Aristarchus' heliocentric theory. Posidonius approved of Seleucus' lunar-tidal views but rejected his heliocentric position; in so doing, he had much to do with the prevalence of the former view and the repudiation of the latter in subsequent ages.

A digest of the contents of Posidonius' scientific writings, if it were possible to prepare one, would serve as an ideal introduction to Roman theoretical science; but since none of his works has survived, let us turn instead to three handbooks which largely represent Posidonian science and at the same time afford innumerable parallels and correspondences with the works of Roman writers on science. The reader should bear in mind that the first two of the three are of uncertain date and the third was compiled early in the second century of the Christian Era. Nevertheless, they reflect the state of science not of the period in which they were written but of the age between Hipparchus (c. 192–120 B.C.) and Posidonius, and for that reason are most appropriately discussed in this place.

Many controversies have arisen among scholars about the identity of the first author, Geminus, and the true nature of his *Introduction to the Phaenomena*. The book in its present form appears to be either a late abridgment of the original work by Geminus or the original work in a deteriorated state as a result of additions or revisions introduced by later compilers. Doubts about the identity and date of the author have arisen from the facts that a statement in the sixth chapter refers to either 77 or 137 B.C., and that views are presented contrary to those that Posidonius is known to have held. For these reasons some scholars have assigned the author to the second century B.C., thus precluding the possibility of Posidonian influence.[4] In fact, it is generally agreed that Geminus' book represents the astronomy of Hipparchus' time or of the period just before him better than that of Posidonius' time. Despite these cogent arguments the usually accepted view is that the Geminus whose name appears with the title of the *Introduction* is the same one who is known to have written a commentary (c. 70 B.C.) on Posidonius' *Peri meteoron*. The Geminus of the commentary was a resident of Rhodes, a pupil of Posidonius, and a Stoic philosopher.[5] That some of the views of the *Introduction* differ from those of Posidonius is natural enough; handbook authors were compilers, who were not so much concerned with consistency as with recording all opinions that they regarded as authoritative upon a subject. The respect they evince for authorities as early as Pythagoras and Plato, mingling the views of these men with those of contemporaries and imputing to early masters views developed later and discussed in commentaries on their works, indicates that inconsistencies should not be considered crucial in determining the date of a work.

Whatever vicissitudes the book may have suffered before reaching us in its present form, Geminus' *Introduction* would have served the readers of the time as a very satisfactory manual of positional astronomy and mathematical geography. It is written in a clear style and exhibits sound although rather elementary comprehension of the subject. The book is slightly over a hundred pages long, is divided into eighteen chapters, and treats topics in a conventional way, without ingenuity or originality.

It opens with a discussion of the zodiac and explanation of the inequalities of the quadrants of the sun's orbit—the figures for the number of days in the four seasons being $94\frac{1}{2}$, $92\frac{1}{2}$, $88\frac{1}{8}$, and $90\frac{1}{8}$. The Stoic view is expressed that the stars are not fixed in a celestial sphere but are set farther from or nearer to us. Geminus then gives the periods of planetary revolutions, including that of the moon. Next follow an elaborate treatment of the signs of the zodiac, a brief discussion of the constellations, and short descriptions of the celestial axis and poles and of each of the celestial circles. Geminus' list includes, in addition to the celestial equatorial, tropic, and arctic circles, the colures, zodiac, ecliptic, horizon, meridian, and Milky Way—a longer list than is found in most handbooks. Next he takes up the varying lengths of day in different latitudes and the amount of time the signs of the zodiac take to appear fully.

After the digression of chapter 8, which has provided specialists on ancient chronology with one of their most valuable sources of information about various luni-solar cycles used in early Greek calendars, the book returns to conventional handbook topics: an account of the moon's phases; an explanation of solar and lunar eclipses and of the peculiar motions of the sun, moon, and planets; a chapter defining the technical terms used by writers on astronomy in referring to the risings and settings of stars; another chapter on the circles described by the fixed stars; and then a discussion of mathematical and physical geography, including a presentation of Cratean views about human habitations in other quarters of the globe. That Geminus preferred Eratosthenes' estimate of the earth's dimensions to that arrived at by Posidonius does not necessarily indicate that Posidonius was not one of his chief authorities. Posidonius is known to have adopted Eratosthenes' figure before he discarded it for his own estimates.

The second of the Greek handbooks representing the Posidonian tradition was compiled by Cleomedes, a man about whose life almost nothing is known. Wide differences of opinion exist regarding his dates: some scholars place him after Posidonius in the the first century B.C., others in the first or second century of the Christian Era.[6] Cleomedes' handbook, *On the Cyclic Motions of the Celestial Bodies*, is distinctly inferior to Geminus' treatise and yet contains some valuable information. It is generally agreed to be mainly a digest of Posidonian information, but it does not exhibit a compiler's customary sense of proportion in handling routine topics. The treatment of the planets is very scanty: a brief paragraph in Book I gives the order of their distances from the earth and their sidereal periods; another in Book II gives the deviations of their orbits from the ecliptic and their periods of the synodic revolutions.

Book I opens with arguments to prove that the universe is spherical, surrounded by void on all sides, that the earth is at the center of the celestial sphere and hence at the bottom of the universe. The second chapter deals with the five celestial parallels, the zones, and the earth's habitations; it describes the latter as diametrically or transversely opposed to each other, a view derived from Crates by way of Posidonius. The third and fourth chapters are brief, explaining the diurnal motion of the celestial sphere, the independent motions of the planets, the movement of the sun through the zodiac, and the resulting effect of climate on earth. The next two chapters deal with the effect of the inclination of the axis of the universe upon the length of days and nights in the northern and southern hemispheres and the effect of the eccentricity of the sun's orbit upon the length of days, nights, and seasons; they include Posidonius' arguments in support of a habitable torrid zone. An entire chapter on the earth's habitations follows; and the remaining chapters of Book I contain further arguments that the universe and the earth are spherical, that the earth is at the center of the universe, and that in comparison with the universe it is a mere point—all common-place features in handbooks on astronomy. The tenth chapter, however, offers the only detailed account that has come down to us of the procedures used by Eratosthenes and Posidonius to determine the earth's circumference.

Book II opens with a lengthy chapter, derived from Posidonius, which in rather bitter tones undertakes to refute the notion of Epicureans that the sun is only as large as it appears. The refutation includes an account of Posidonius' calculations of the size and distance of the sun (diameter $39\frac{1}{4}$ times as large as the earth's; orbit 10,000 times the earth's circumference) and Hipparchus' estimate of the sun's size (1,050 times the earth's size), a figure which obviously refers to volume and not diameter; it also includes his own rather inept calculations of the sizes and distances of sun and moon. Disregarding the results of these computations, he concludes from other arguments that the sun is larger than the earth. Then follow chapters on the sizes of the planets and stars (it is conceivable that some stars surpass the sun in size, he states), the phases of the moon, the causes of the phases and of eclipses, and a brief chapter on planetary deviations and synodic periods.

Cleomedes' manual has been noticed in histories of early science mainly because of two features: the detailed accounts, just referred to, of Eratosthenes' and Posidonius' calculations on the circumference of the earth, and Posidonius' method of calculating the sun's size and dimensions; and Cleomedes' explanation of the phenomenon of so-called "paradoxical" eclipses. Eratosthenes' and Posidonius' procedures are discussed fully in histories of mathematics and ancient science. It is enough to recapitulate here that both men based their calculations of the earth's measurements upon observations of declinations—Eratosthenes of the sun, Posidonius of the star Canopus; from estimates of the distance between points of observation they were able, by geometrical methods, to arrive at rough estimates of the earth's circumference. Cleomedes' discussion of paradoxical eclipses (an eclipsed moon on the eastern horizon while the sun is still visible on the western horizon) is noteworthy because it suggests the correct explanation, namely, atmospheric refraction. It is hard to credit him with an original observation here, as some historians of science are inclined to do;[7] throughout the book he is reporting received opinions, and on occasion gross misapprehensions, such as the one about all solar eclipses being annular.

A *Manual of Mathematical Knowledge Useful for an Understanding of Plato,* by Theon of Smyrna, the third and last of the Greek handbooks in the Posidonian tradition to be discussed here, is an immensely valuable work to anyone surveying this field—a veritable storehouse of correspondences with the doctrines and data contained in other Greek and Latin handbooks. It is through such omnipresent correspondences, even better than through Theon's scrupulous citations of authorities, that we realize how stereotyped is the material purveyed by the compilers of these manuals. In some handbooks on arithmology the correspondences range from precise data or figures set down in tabular or reduced form to passages running to six pages of verbatim copying or bald translation. Large-scale appropriation of material from works of predecessors, usually made with no acknowledgment of the source or else with false citation, was a common practice among Greek writers on Pythagorean arithmetic and arithmology and among Latin compilers in general. It is not without significance that our word *compiler* comes from the Latin verb meaning "to plunder." In recent decades scholars have arranged side by side the texts of over a dozen Neopythagorean or Platonic handbooks or commentaries to show the sort of extended copying and translating these professional hacks engaged in.[8] In several Latin handbooks nearly every statement after the author's introduction has parallels in the works of other authors.

Theon's *Manual* affords parallels in a number of fields because, even though it is not, strictly speaking, a *Timaeus* commentary, it follows the pattern of such works in covering the entire universe, as Plato did in his original work. Plato's discussion, in the *Timaeus* and in the closing episode of the *Republic,* of the planetary spheres and the harmonies produced by their motions, and his use of Pythagorean arithmetic in the Creator's fabrication of the World-Soul, gave occasion to a host of commentators on his works to expound on the intricacies of Pythagorean arithmetic, geometry, musical theory, and astronomy. In mathematics and musical theory the commentators reproduce with little change traditional doctrines that originated in the early Pythagorean school and might therefore represent Plato's mathematical background. But in astronomy they show no regard for chronology; they credit Plato with demonstrations and discoveries that came two or three centuries later. Because the correspondences in handbooks are most abundant in fields handled by *Timaeus* commentators, it has generally been acknowledged that the dominant tradition in ancient popular science is represented by a nine-hundred-year line of *Timaeus* commentaries, beginning shortly after Plato's death.

That Posidonius wrote a commentary on the *Timaeus*, keystone in an arch of Greek and Roman commentaries erected by earlier scholars, seems sufficiently attested in ancient writers and is indicated by scores of bits of evidence that are hard to explain on another basis. But a number of recent scholars, reacting to the prevailing tendency, have adopted a rigorous attitude of demanding absolute evidence, and a few deny that such a commentary ever existed.[9] Such differences among scholars show how problematical are the bases of argument in attempts to trace sources in an area where most of the works have been lost and citations, when trustworthy, are usually by author alone, not by author and title; in the case of Latin compilers they are almost never trustworthy.

Theon's *Manual* was probably written in the first half of the second century of the Christian Era, but since it mostly represents a state of knowledge existing in the age of Posidonius and even earlier, it is appropriate to consider it here, among the works that helped to mold the form and content of Roman theoretical science. Our information about the author is derived from his book. To assume that this Theon is the same man as Theon "the mathematician," whose observations of Venus and Mercury in the years 127–132 are recorded by Ptolemy in his *Almagest*, would be precarious, for the Theon we are discussing exhibits the characteristics of an uncritical compiler gathering information, good and bad, from earlier authorities.[10] The book has much to commend it, chiefly because it preserves a large amount of Hellenistic astronomical theory that would otherwise have been lost. Theon also makes gracious acknowledgment of his sources, citations which have not been impugned by later scholars. The citations average nearly one for each of the two hundred pages of text. Theon's chief authority was Adrastus, a contemporary Peripatetic philosopher who wrote commentaries on some works of Aristotle and Theophrastus and on the *Timaeus*. It was Adrastus' lost *Timaeus* commentary, quoted by Proclus, which was in all likelihood the source of Theon's frequent and extended quotations. Adrastus is believed to have drawn his work, in turn, largely from Posidonius' commentary on the *Timaeus*.

Theon's opening paragraph states that his *Manual* has been designed to give readers who have not had training in mathematics the essentials for comprehending the five branches of Plato's mathematics: arithmetic, geometry, stereometry, musical theory, and astronomy. The author, however, gives no formal treatment to geometry and stereometry, assuming that his readers are already familiar with the elements; and his discussion of music is largely devoted to proportions and means and the ratios existing in the harmonious universe. Theon's book was formerly regarded as a fragment—the portions on arithmetic and astronomy surviving, the other mathematical disciplines being lost—but a more recent view held by the latest editor, Hiller, and others is that the work is virtually complete.

The section on Pythagorean arithmetic and the mathematics involved in Pythagorean musical theory and Platonic harmony of the spheres occupies more than half the book. Theon's discussion proves to be an orthodox treatment, closely resembling the material in numerous other manuals of arithmetic and harmony, and, in particular, Nicomachus' *Introduction to Arithmetic*. The detailed comparison of the texts of Theon and Nicomachus made by F. E. Robbins [11]—which will be mentioned again in connection with Nicomachus—reveals marked similarity of content throughout but almost no verbal similarity. The natural conclusion is that both works were derived from an ultimate common store of materials, a corpus of traditional Pythagorean doctrines, but that their immediate and perhaps intermediate sources were different. The revival of Pythagoreanism under the early Roman Empire was probably responsible for the publication of the works of Theon and Nicomachus. Theon drew his mathematics and harmonic theory largely from Adrastus, who, in turn, is supposed to have depended mainly upon Posidonius. Numerous bits of admittedly tenuous evidence such as this lead us to suspect that Posidonius may have had more to do with the revival of Pythagoreanism than anyone else.

Theon's section on astronomy was also largely derived from Adrastus, with extended quotations from Dercyllides, who lived during the reign of Augustus and wrote a commentary on the myth in Plato's *Republic*; and indirectly from such well-known astronomers as Eudoxus, Eratosthenes, Archimedes, and Hipparchus. Its early chapters offer a conventional treatment of mathematical geography: proof of the earth's sphericity (Theon uses the argument of ships disappearing over the horizon, in addition to the earlier arguments used by Aristotle); the small size of mountains compared to the earth's dimensions; the earth's circumference; the celestial circles; maximum deviations of the planets from the ecliptic (maximum deviation for the sun is given first as one degree, later as half a degree, both figures which crop up in the handbook tradition); the discrepant orders of the planets according to some [early] Pythagoreans and other [late] Pythagoreans; and the harmonic intervals to be assigned to the seven planetary spheres, the celestial sphere, and the earth. The later chapters, on the direct, stationary, and retrograde phases of the planets' orbits, are more interesting and valuable, preserving for us a number of theories about eccentrics and epicycles advanced by Hipparchus and other eminent Hellenistic astronomers. We find epicyclic motions assigned to Venus and Mercury, theories later appropriated from Theon by Chalcidius and credited to Heraclides of Pontus, although epicycles had not yet been developed by the time of Heraclides, who was regarding Venus and Mercury simply as satellites of the sun. We even find epicyclic theories attributed to Plato. There is a lengthy, cryptic passage propounding the Stoic doctrine, perhaps influenced by Aristarchus' heliocentric hypothesis or by early Pythagorean speculations, that the sun is the heart and center of the *animate* universe though the earth is the *mathematical* center; this doctrine was echoed by Pliny, Plutarch, Chalcidius, Macrobius, and Martianus Capella, and through them was transmitted to the Middle Ages. Since the theories of Heraclides, Aristarchus, and Hipparchus have been fully discussed in histories of early astronomy, they will not be dwelt on here.

What historians of science have not fully appreciated is that handbooks like Theon's are merely repositories of received opinions or of theories of esteemed authorities. The compiler is respectful of the written word and of the reputation of his renowned predecessors. To be on the safe side he includes many conflicting theories and data. He is not competent to decide where the truth lies; in fact he frequently does not seem to be aware that the views expressed are contradictory. Occasionally he puts on airs and criticizes an earlier compiler for admitting contrary opinions: Theon points out that one of his authorities says the heptachord is the image of the world and a moment later uses a scale of nine strings.[12] Actually very few compilers were competent to handle the technical matters involved in the musical intervals producing the harmony of the spheres, and the handbooks are full of discrepant statements on this subject. Theon himself presents two variant but fixed orders of the planets, the theory of epicyclic motions for Venus and Mercury about the sun, and the Stoic doctrine of the sun as the heart of the universe—each wholly or partially incompatible with the others. It would therefore be a mistake to draw sweeping conclusions from any isolated statement or theory in such a mélange of doctrines. One should not suppose, because Theon adopts Eratosthenes' figure for the circumference of the earth, that he is not under Posidonius' influence elsewhere, or assume, because Theon assigns a slight deviation from the ecliptic to the sun when Hipparchus had demonstrated that the sun does not deviate, that Theon's astronomy must be pre-Hipparchan. Theon quotes extended passages from Hipparchus elsewhere, and is himself confused about the sun's deviation, giving two discrepant figures.

Before turning our attention to Roman science, we should make some pointed observations about the man who is our most important source of information on Greek geography. Strabo (c. 63 B.C.–A.D. 21) did not enjoy as eminent a reputation as a geographer as did Ptolemy, nor did his attainments as a scientist compare with Ptolemy's; yet he had a grand conception of geography, embracing the four major divisions—mathematical, physical, political, and historical—and produced a work which, because of its comprehensive scope and attractive presentation, is by far the most valuable geographical treatise to survive from antiquity. One would hardly suspect Strabo's *Geography* of conforming to ancient handbook traditions; its style is much too argumentative and lively for a handbook. Mere bulk would also exclude it: eight volumes in English translation in the Loeb Classical Library. Most important consideration of all, it did not belong to the tradition in antiquity and was either ignored by or unknown to compilers. Strabo himself, however, does exhibit many traits characteristic of compilers and, as we shall see, covertly incorporated a conventional handbook treatise on geography into his vast work.

Strabo was born at Amasia in the distant province of Pontus, on the Black Sea, yet, like the Greek intellectuals Polybius and Posidonius, whom he greatly admired and who were his leading authorities, he came to be an intimate friend and escort of prominent Romans. He traveled widely to distant capitals with Roman administrators and men of affairs, evidently more as a congenial companion than in the interests of geographical research. He bubbles with admiration for Roman officials and the efficiency of their administrative organization, and regards the Roman Empire as a universal boon to mankind. Like Polybius and Posidonius, he turned his travels to good advantage in the preparation of his *Geography*, which he indicates was written to make available to statesmen and administrators a knowledge of all the countries of the empire. Ironically, Roman writers seem to have had no knowledge of the book's existence. Pliny does not list Strabo in his vast array of authorities. A reasonable explanation has been offered that Strabo retired to Pontus to write his work, after collecting his materials at Alexandria and Rome, and then committed the folly of publishing his volumes in his native land.[13] There is a Byzantine tradition of Strabo but none in Latin before the Renaissance.

It is obvious from Strabo's opening remarks that he does not conceive of geography as a specialized and technical discipline; rather it is a subject suited to polymaths because it embraces a knowledge of all things in the heavens and on earth. Geography is thus akin to philosophy and, like it, provides its students with insights for the pursuit of successful and happy living. Strabo shows an obvious distaste for mathematical geography; yet because the practice of using a summary of geodetic and astronomical principles to introduce a general work on geography was firmly established, and because his leading authority, Posidonius, set an immediate precedent, he includes an extended discussion of the subject in Book II. Though he lacks competence to handle mathematics, Strabo has the temerity to criticize Eratosthenes and Hipparchus. His travels were far-reaching, yet restricted in many areas, and he admits that he got no farther west than Etruria in northern Italy. For Gaul and Spain he depended upon literary sources—Posidonius, Polybius, and Caesar. In all of Greece, surprisingly, he touched down only at Corinth. It is in Greece that we best observe the deficiencies of his methods of research. He let his veneration of Homer lead him astray and, instead of obtaining information from some up-to-date geographical work, used Apollodorus' commentary on Homer's Catalogue of Ships. To Strabo Homer was a philosopher and his poetry a "philosophical production"; he takes issue with Eratosthenes, who felt that Homer should not be regarded as a reliable authority for factual research.

Because of his lack of qualification, Strabo was unable to produce the mathematical introduction that would have befitted his large work in seventeen books. Instead he devotes most of the first two books to an interpretation of the geographical references in Homer and to a critique of Eratosthenes' treatise on geography, with briefer discussion of geographical matters found in the writings of Crates, Hipparchus, Polybius, and Posidonius. Then, just before embarking upon his regional description of the known world, Strabo inserts a separate treatise that conforms with the conventions for compiling a succinct and comprehensive layman's handbook on geography. In one extended chapter (II.5) he surveys the entire subject, beginning with a discussion of the shape and zones of the earth and heavens, the size and position of the earth in relation to the universe, the extent of human habitations, and the shape of the inhabited world. Next he proceeds, in the manner of a cursory coastal survey, to sketch the known inhabited world, girt by an ocean which invades it in four great gulfs—the Caspian, Persian, Arabian, and Mediterranean. He concludes with a discussion of latitudes and of the equinoctial hours of daylight recorded at the various parallels on the longest day of the year.

Thus Strabo's bulky work consists of three distinct parts: two of them are introductory, and the body of the work is a regional geography. The first introduction follows the approach of philosophers and historians who regarded Homer as an important authority on geography; the second presents the standard information that is found in handbooks or in the geographical portions of encyclopedic works.

PART TWO

ROMAN SCIENCE

OF THE REPUBLIC

AND THE

WESTERN EMPIRE

V

LATE REPUBLICAN TIMES

Anyone who examines the civilizations of the Greeks and Romans is immediately impressed with the striking differences between the mentalities of the two peoples. He might be led to conclude that their temperaments were antithetical—the Greeks theoretical and intellectual, the Romans practical and anti-intellectual. Such a generalization would of course be unwarranted, but it probably comes closer to the truth in their attitudes toward science than in any other respect. The Greeks manifested a strong aversion to applied science; they called it *banausikon*, meaning "fit for mechanics."[1] The Romans, on the other hand, found it hard to absorb any small amount of theoretical science, no matter how digestible the form in which it was presented. An observation made by Joseph Partsch, a nineteenth-century historian of geography, illustrating these differences, is so apt that it has been used repeatedly by historians of science: the Greek scientist was typified by Eratosthenes, who undertook to measure the circumference of the globe by astronomical observations; the Roman scientist by Marcus Agrippa, who measured the length and breadth of each province of the Roman Empire by computing the distances recorded on milestones along imperial highways.[2]

In spite of this difference in attitude toward pure and applied science between the Greeks and the Romans, in the realm of handbook science we find surprisingly strong bonds of sympathy and common motivation. Greek manuals on scientific subjects were written concisely and clearly, with no literary pretensions, to provide educated laymen with a quick grasp of a discipline. Such handbooks were typical products of the later Hellenistic Age, when the masters of rival philosophical schools were embracing each other's tenets, historians were stressing the importance of geography, geographers were claiming all branches of knowledge as pertinent to their discipline, and the goal of learning was a happy and successful life. Greek scientists of the second and first centuries were still continuing their researches and observations at Eastern centers of science and learning, but not with the fervor and brilliance of the earlier Alexandrian investigators. Scientific studies at the Alexandrian Museum and Library had fallen into a decline after their short-lived flowering in the third century. The Romans made their first large-scale contacts with Greek society at a time when the most highly respected Greek intellectuals were cosmopolites and polymaths. At no stage in their intellectual history could the Greeks have been more attractive to the Romans.

When cultivated Greeks first began to spellbind the Roman nobility and nouveaux riches, they must have found their own handbooks ideally suited for the purpose. One might almost suppose that their textbooks had originally been designed to be translated or paraphrased into Latin, as well as for Greek readers. A Roman gentleman was willing to dip into Greek abstract disciplines—if it was becoming fashionable to do so—but he wanted only the bare essentials, with no complexities or waste of effort.

When the handbook fashion caught on at Rome, it surpassed all other literary productivity in its volume and popular appeal—in spite of the fact that alongside this mass of encyclopedic and didactic writing there emerged some of the world's most inspired verse and most vigorous and incisive prose. It is noteworthy that the best period in Latin literature coincides with the time when the Romans showed their greatest interest in Greek theoretical subjects; but that period was confined to two centuries during the late Republic and early Empire.

The attraction that books of the *vade mecum* type had for Roman readers is evidenced in many ways. Titles of hundreds of lost Latin manuals and collections of practical information are known, and over a hundred Latin handbooks have survived in toto or substantial part, or are represented by enough fragments to indicate their scope and character. Practicality determined the popularity of the handbooks: the most numerous titles are found in such fields as agriculture, the military arts, law, and rhetoric. Titles have also been found on almost every conceivable subject in which Roman readers might have had an interest: pharmacology, toxicology, metrology, surveying, dream lore, gem lore, and all sorts of divination; books for scholars and specialists on philology, orthography, and numerous topics of antiquarian interest; and manuals for all the professions.

Further evidence of the Roman penchant for handbooks is revealed in their keen interest in Greek didactic works of every age and type. Greek poets like Hesiod, Empedocles, and Aratus had demonstrated that didactic subjects could be handled successfully in verse; in fact, the vogue these writings enjoyed may be attributed in no small measure to their verse form. The Romans too felt this attraction. Poetasters tested their abilities by translating such poems into Latin verse, assured of an appreciative response to the subject matter if not to the quality of the verse. Versifiers were encouraged to create didactic poems on Greek models. Judging from the titles and the reputations of the Latin poets, it is hardly to be regretted that most of their efforts have perished. At the same time, some of the finest Latin verse was also lavished upon didactic subjects: in Lucretius' *On the Nature of the Universe* (following Empedocles and Epicurus), in Virgil's *Georgics* (following Hesiod and Aratus), and in Horace's *Art of Poetry*. These superb poems are, in a sense, handbooks of their subjects. Expository treatises on poetic composition by Neoptolemus of Parion (third century B.C.) and Philodemus (c. 110–38 B.C.) have been regarded as the leading sources of Horace's work. Fragments of Philodemus' treatise were recovered among literary papyri found at Herculaneum.

Another popular type of book which was closely related to the Roman handbook genre, though it did not actually belong to it, was the abridgment or epitome. Romans of the upper class were usually too busy or too responsive to other distractions to wade through voluminous works; thus they created a demand for all sorts of *breviaria*, drastic reductions that seldom satisfy one's curiosity about the contents and literary quality of the works they epitomize. Abridgments also reduced the cost of reproducing a manuscript running to many scrolls. As in Greece commentaries on famous works often superseded the works they were expounding, in Rome abridgments and epitomes of abridgments, like those of Livy's *History of Rome*, consigned bulky original works to oblivion. By the third and fourth centuries of the Christian Era, as we shall see, such abridgments began to have an important effect upon the Latin handbook tradition.

During the same period Latin commentaries also began to assume a prominent place in handbook science. By that time commentaries had gained a position in Latin letters comparable to that they held in Hellenistic Greece six centuries earlier. Two noteworthy Latin examples which will be discussed later are Chalcidius' translation of two-thirds of Plato's *Timaeus*, together with his commentary on that portion, and Macrobius' *Commentary on Cicero's Dream of Scipio*. Chalcidius' work made Plato's fascinating *Timaeus* available to Latin readers and revealed to them the sort of agile ratiocinations that could spring from it. Both commentaries became basic manuals of medieval Latin science and Platonism, and held a dominant position until well into the thirteenth century.

Latin handbooks closely reflected Roman attitudes toward education. Before the period of Greek influence, education was focused upon agriculture, the arts of warfare, law (learning the Laws of the Twelve Tables by heart), and medicine, with the aim of producing prosperous landowners and a militarily powerful nation. To the head of a household the preservation of the health of his family, his slaves, and his livestock was a primary concern; and so we find, in treatises on agriculture by Cato, Varro, Columella, and Palladius, sections on the treatment of human complaints and on the diseases of animals. There were also numerous separate works on the treatment of human disease and on veterinary medicine—shocking armamentaria of incantations, coprophagy, and shotgun prescriptions reputed to be efficacious for man and beast, that bore no resemblance to Greek scientific treatises on nosology and therapeutics. It is widely presumed that a main source of many of the Roman manuals on agriculture was a Carthaginian work by Mago, translated into Latin by order of the Roman Senate after the destruction of Carthage.

Later on in the Republican period, as influential Romans became acquainted with Greek learning through the visits of Greek intellectuals, having educated Greeks as slaves in homes in Italy, and through various cultural exchanges resulting from the Roman conquest of such important intellectual centers as Athens, Pergamum, Rhodes, and Alexandria, they were gradually won over to endorse a Greek style of education —but seldom enthusiastically and nearly always with a disposition to adapt the Greek practices to Roman daily needs. Little effort was required to persuade Romans of the practical virtues in the traditional Greek training in rhetoric. The development of oratorical skills was always considered an important part of the training of a Roman politician. Back in the fourth century B.C. in Greece Isocrates, too, had argued the direct relation of rhetorical training to the development of future statesmen when he was trying to establish the position of rhetorical studies in the curriculum of Athenian schools. In Isocrates' day the academic disciplines were known as *technai*, which suggested their practical virtues; when the Romans introduced Greek disciplines, they translated the term as *artes*. Thus our designation "liberal arts" is derived from an attitude that liberal studies are practical in their functions as preparatory subjects.

With the Greek mathematical quadrivium, however, it was an entirely different matter. Hard-headed Roman parents could not imagine how abstract mathematics, theoretical astronomy, and harmonic theory could contribute anything to a lad's grooming for administrative posts or vigorous participation in empire building. Greek household tutors argued, as had Hippias and Isocrates centuries before, that mathematics sharpened wits; and Polybius pointed out to his aristocratic hosts that a general could use a knowledge of astronomy in moving legions from place to place. Greek arguments prevailed, at least during the period of keenest intellectual stimulation: abstract mathematical studies were widely introduced into Roman schools, as writers testify who recall the tedium of their lessons in arithmetic and geometry. But we may suspect that such theoretical studies were tolerated at Rome not so much for their intrinsic virtues as because it had become fashionable for Romans to retain tutors to educate their sons in the Greek manner. Other reminiscences in Latin writings relate amusing occasions—found also, to be sure, in Greek literature—when a realistic parent interrogated a son about his mathematical studies and wondered what application they could have for the market place or household management. The meaningful difference is that the sympathies of Greek readers lay with the son, those of Roman readers with the father. Occasionally a Roman gentleman might decry the constant harping upon practical subjects, as Horace does in the *Art of Poetry*; but the position of pure mathematics was always a precarious one at Rome—at first a dogged struggle to gain it recognition in the curriculum, then a period of fashionable toleration, and finally, during the empire, a gradually deteriorating position in the schools.

Roman patricians were not averse to subjects that would sharpen young men's wits. If they denied that such virtues existed in mathematics, they felt that they were present in philosophical studies. Metaphysical philosophy, as the Greeks had developed it, could not be expected to appeal to the dilettante instincts of Romans, but they did appreciate that merely following the rigor of Stoic argumentation might develop a youth's powers of logic, and that the suspended judgment dear to practitioners of the New Academy school might prove useful to young jurists training to anticipate both sides of an argument. Furthermore, any attention that Greek philosophers gave to morality in their writings of course commended them to a patrician society which outwardly took its standards of morality very seriously.

How the Romans turned the Greek amenities and urbanities to practical advantage may be seen in the performances of Greek dramas at Rome. The plays of the classical masters, in Latin translations and adaptations, enjoyed a vogue among the fashionable set. But aspiring orators also attended the theater to study the declamatory techniques of tragic actors. The young Cicero avidly marked the postures, gesticulations, and voice control of the famous actor Roscius, dreaming of the day when, as a politician, he could emulate such performances in the Forum and Senate house.

Handbooks became the bridge over which the various Greek disciplines were brought to Rome. The most expeditious way to make the subject matter of a discipline available to Latin readers was to have a translation of a handbook prepared. We observe, from instances in which we have a Greek original to compare with a Latin translation or adaptation, that the translations were quite free where the subject matter was difficult, omitting or paraphrasing complicated discussions and introducing numerous examples to assist the reader's comprehension. Except in their occasional and sometimes free use of examples, the Latin copies were completely dependent upon their models and showed scant ability to make allowances for a new set of conditions. For example, the formal study of grammar got a late start in Greece; it was not until near the end of the second century B.C. that Dionysius Thrax produced the first Greek grammar. This book became a model for the early Latin treatises on grammar prepared by Varro and Palaemon, and continued to be followed so slavishly by Latin grammarians who wrote near the fall of the western empire that we find discussions of the definite article, though none exists in Latin, and differentiations between the subjunctive and optative moods, though Latin employs only subjunctive forms. Since English grammarians based their books on late Latin grammars, Dionysius may be regarded as the founder of an ossified tradition that has endured to the present and has always been too dependent upon earlier concepts to be considered a sound descriptive science.

Hellenistic treatises on both applied and theoretical subjects were imported and adapted to Roman educational needs. Some of the more popular works, like Aratus' *Phaenomena*, were translated into Latin repeatedly and became standard textbooks in the schools. An insight into the sort of reception that poem found among Roman educators is offered by Quintilian (*fl.* A.D. 85), who was recognized as the leading literary critic and authority on education of his time. Quintilian does not consider the style of Aratus' book at all attractive, yet he acknowledges that a grasp of astronomy is necessary for an understanding of poetry. Another equally successful poet, Nicander, known as a *metaphrastēs*, or professional translator, produced several metrical versions of practical handbooks. His poems on farming and bee keeping were familiar to Cicero and Virgil. It is not within the scope of this volume to discuss the multitude of Greek technical treatises that were appropriated and copied in Rome. One more instance will suffice. It used to be supposed that Tiro, Cicero's amanuensis, was the inventor of shorthand. It is now understood, from specimens of shorthand and fragments of manuals on tachygraphy recently found in papyri, that the Greeks made considerable progress in this field, and it is considered likely that Tiro derived his system from some Greek manual.[3]

Among the Greeks the lay handbooks represented a low order of science, but at Rome there was only one level of scientific knowledge— the handbook level. Even the most intellectually curious Romans, like Lucretius, Cicero, Seneca, and Pliny, were satisfied to obtain their

knowledge of Greek science from manuals and made no original contributions. Latin handbook science was outdated from the beginning,
since it was a synthesis of Greek investigations and theories that were
a century, or even two or three centuries, old by the time they had
been introduced at Rome. And because of most Latin compilers' lack
of aptitude for theoretical studies, handbook traditions of Greek science
suffered further deterioration each time they passed through the hands
of a new compiler. Cicero's attitude illustrates a Roman intellectual's
feelings about theoretical science. He expresses gratification, near the
opening of the *Tusculan Disputations,* that whereas the Greeks exalt
the position of pure geometry the Romans sensibly apply the study to
mensuration and reckoning.

The Latin compilers hoped that a display of scholarship would conceal their lack of professional competence. As we shall see, they were
aware of the reputations of eminent Greek scientists and pretended to
be using their works as sources of information; but in the majority of
cases the immediate source of a new Latin compilation, during the Republican period, was some Greek handbook. The standard practice of
Roman compilers was to cite as their own authorities the names that
the Greek compilers acknowledged as their sources. Thus, at one
stroke, they gained greater authority for their own compilations and
concealed their extensive appropriations of already assimilated materials. Many scholars in the past have been deceived by citations of the
works of Eudoxus, Eratosthenes, Archimedes, Hipparchus, and Ptolemy. Like the numerous citations of Pythagoras, who committed nothing to writing, these references must be flatly repudiated. Parts of
Euclid's *Elements* were translated into Latin by Boethius early in the
sixth century, it is true, and Roman schoolboys studied digests and
extracts of Euclid's book during the classical period; but it would have
been a phenomenon for a Roman to make a serious effort to comprehend the theoretical works of Archimedes, Hipparchus, or Ptolemy,
and there is no conclusive evidence of success in such an effort.

Most of the Latin handbooks of which we have tangible knowledge
do not come within the scope of this volume since they deal with practical subjects. Manuals belonging to the early period, before Greek influence had become effective, would naturally be wholly practical. A brief
consideration of one of these treatises—the earliest extant book on

Roman science—will serve as an appropriate beginning of our study and at the same time throw some light on Roman attitudes toward theoretical matters.

The first Latin encyclopedist was the unforgettable Cato the Elder (234–149 B.C.). Born of hardy plebeian stock, Cato grew up on a Sabine farm, distinguished himself in military campaigns, engaged successfully in politics, attaining the consulship in 195, and soon after settled down in the Senate to become an arch critic of the effete new manners being adopted by the aristocratic set of his day. Patriot and vehement jingoist, Cato reacted fiercely to the invasion of all foreign influences and particularly to the threats posed by Greek intellectuals against traditional Roman discipline and modes of living. He set himself resolutely to counteract the influence of Hellenophiles like the Scipios and the growing fashion of entertaining or retaining Greek philosophers and tutors in Roman households. In his later years Cato bitterly resented the influx of Greek immigrants who were beguiling Roman youth with their lectures in the Forum or being lionized in the salons of the nobility. He urged before the Senate the expulsion of the three celebrated Athenian ambassadors, Carneades, Diogenes, and Critolaus, who were using the occasion of a protracted delay in Rome to lecture on philosophy.

Cato freely expressed his contempt for literature yet was himself a prolific writer; this was for him one way of combating Greek importations. All the topics he discussed were practical; and because of the wide range of his interests, his matter-of-fact approach, and his master foreman's style of writing, he was regarded by staid and homely Romans as an authority in all fields of knowledge and his books were cherished as a truly Roman heritage. He left a series of manuals of instruction, addressed to his son, on agriculture, health, military tactics, oratory, and law. Only the handbook *On Agriculture* has survived, a collection of rustic precepts on the complete farming operation. There is hardly a sentence in the book that does not contain an instruction, command, recipe, or apothegm. Every page reflects his detestation of the littérateurs of the Scipionic Circle by making it appear that a respectable Roman has no time for philosophical reflection on life when the business of successful living is a full-time active occupation. Page after page of his stacatto commands, of tasks to be performed from sunup to sundown in every season of the year, cover all the routines of overseeing a large estate. Instructions on grafting and pruning, reminders about cleaning vessels and repairing implements, recipes for the kitchen, prescriptions for ailing cattle and slaves, helpful hints about buying farm equipment, marketing produce, and caring for the watchdog come all jumbled together, jotted down like reminders on an agenda pad.

Cato is often regarded as one of the unspoiled "old Romans," a representative of the ancestral type of men of hallowed memory. This anachronistic conception stems largely from the idealized biographical sketches of him, by Plutarch, and by Cicero in his attractive essay *On Old Age*. Cato's political career was a realistic and opportunistic one; but his one-man campaign to thwart the invasion of Greek literature with a set of vernacular manuals was most unrealistic. If there was to be any Latin literature, it would have to be almost wholly derived from Greek models. Some indication of how unsuccessful Cato's campaign was, even in the handbook field, may be seen by contrasting his avoidance of literary authorities with the great respect shown for earlier authors by Marcus Terentius Varro, the next important writer in the field of agriculture. Cato's attitude was almost unique; Varro's practice became the standard one for encyclopedists.

Varro (116–27 B.C.) was undoubtedly, as Quintilian said, "the most erudite of all Romans," a man who had familiarized himself with the entire range of Greek encyclopedic studies; Cicero called him "the most voluminous writer." Varro became the prototype for a host of pedestrian Latin scholars who, through reading in many fields and digesting or excerpting the works of earlier authorities in the production of their own compendia, expected to gain a reputation for learning. This type of literary man, if he deserves to be called literary, belonged to the most numerous class of writers in the Latin language. Rome was swept with a tide of manuals, many of them having titles of *artes—ars rhetorica, ars geometrica, ars musicae, ars grammatica, ars gromatica* (surveying), *ars coquinaria* (cooking)—on every sort of topic of instruction. Even Horace called his excellent poem on poetic criticism *Ars poetica*; and when the sensual Ovid entered the field with a poetic parody, the *Ars amatoria*, a manual on love and seduction, sedate Romans must have felt that the gamut had been run. Several of Varro's works, including the two that have survived—*On Agriculture* and *On the Latin Language* (in part)— would be classified as artes.

Varro had many imitators who stood in awe of his reputation, but no one, not even Pliny the Elder, could match his range of reading and his output. Pliny, it must be admitted, was making a determined effort and might eventually have rivaled Varro had he not pursued his researches too close to the slopes of erupting Vesuvius. Cicero, one of the most voracious readers of his day, with an awesome reputation of his own for breadth of knowledge, was filled with admiration for Varro's learning and paid him a gracious tribute in his *Academics* (I.3.9).

At the age of seventy-seven Varro took stock of his work and counted 490 books; but he lived to be nearly ninety and wrote some of the most important ones in his eighties. The grand total has been computed as approximately 620 volumes, counting as one each of the books that made up a larger work. The total number of titles, grouped under the headings literary, historical, and technical, was a mere 74. This is the closest a Roman came to approaching the dubious record of old "bronze-guts" himself, Didymus Chalcenterus, an Alexandrian literary critic who got the epithet from the fact that he was credited with having written between 3,500 and 4,000 books.

Varro, more than anyone else, was responsible for the classical Roman attitude which survived among Latin writers of the Middle Ages: that the man who could assemble the greatest number of authorities on a subject was himself the most reliable authority. Cato mentions no earlier writer in his manual *On Agriculture*; but Varro, in his treatise on the same subject, after a sentimental dedication to his wife and an invocation to the twelve Roman councillor-gods in their respective domains of patronship, proceeds to list fifty Greek authorities whom his wife will find it profitable to consult.

Still greater contrast between Cato and Varro may be seen in their attitudes toward encyclopedic learning. Cato's interests were confined to applied fields, and his treatment of subjects to direct instruction. Varro introduced Roman readers to the Greek type of encyclopedia, compiling a large handbook entitled *The Nine Books of Disciplines*. Though the work has disappeared, much is known about it from later borrowers. The first seven of his disciplines were the Greek rhetorical trivium and the mathematical quadrivium, to which he added, in deference to Roman interests, two books on the practical fields of medicine and architecture. This encyclopedia, alien though it was to Roman ideas of what should constitute book learning, proved to be a signal success. It became one of Varro's most celebrated works, and exerted a weighty influence upon Latin encyclopedic learning in subsequent ages. The Middle Ages rejected his practical fields, as we shall see; but Martianus Capella, by adopting from Varro the seven other disciplines —grammar, dialectic, rhetoric, geometry, arithmetic, astronomy (or astrology), and music—in the fifth century and making them the subjects of his popular handbook, founded the standard and remarkably durable medieval curriculum of the trivium and quadrivium. Ever since, medicine and architecture have been considered fields of professional study.

It is assumed from scanty quotations, rather doubtfully traced borrowings, and faint echoes in later Latin writers, that Varro's encyclopedia was based largely upon Greek handbooks, with heavy indebtedness to Posidonius. Direct quotations in Aulus Gellius' *Attic Nights* indicate that Varro closely followed Greek manuals on Pythagorean arithmetic. Whether he drew his mathematical materials directly from Posidonius or from the same source Posidonius depended on is uncertain, but at any rate correspondences are close; it is supposed that writers like Favonius, Macrobius, and Martianus Capella used Varro as a leading source and that his position among Latin writers was comparable to that held by Posidonius among later Greek writers on arithmetic. Correspondences and resemblances found in Mela's *Chorography* and the geographical portions of the works of Pliny and Martianus Capella are commonly explained by assuming that Varro was a main Latin source of all of them.

A generation or two ago it was widely supposed that Latin science was mainly derived from Posidonius through the intermediary of Varro's works, and that later Greek handbooks were also largely derived from Posidonius through a lost work by Adrastus. More recent studies have shown that a simple and direct line of derivation will not withstand scrutiny from many angles.[4] There is insufficient evidence to give Posidonius and Varro such exclusive positions of authority; and we now know that some writers, like Augustine, Macrobius, Boethius, and Cassiodorus, used Greek sources directly. A precarious attempt was made by Ritschl to reconstruct Varro's *Nine Books of Disciplines* from the handbook Capella wrote more than four centuries later.[5] Tracing sources before and after Posidonius and Varro will always intrigue addicts of *Quellenforschung*, but it must be kept in mind that in an area where surviving remains are scanty, citations of borrowings are unreliable, and the practice of verbatim excerpting or close borrowing is widespread, it is foolhardy to speak dogmatically about the lines of transmission.

Cicero, a contemporary of Varro and his rival as a man of letters, stood in awe of Varro's prodigious scholarship; but unlike Varro he wore his mantle of learning gracefully and lent charm and urbanity to his literary and philosophical discourses. He came to be considered, in the eyes of later Romans and Latin scholars in the Middle Ages, the model of a man of culture and refinement. Cicero was not nearly so attracted to handbook learning as Varro, and there is no indication that he was ever a handbook compiler. His models of style were the philosophers and orators of the golden age of Greece and not the encyclopedists of his own day. He himself appreciated the novelty of his approach and, in writing to his friend Lentulus (*Ad familiares* I.9.23) about his recently completed treatise *On the Orator*, remarks that it does not contain the commonplace rules on rhetoric but embraces the theories of Aristotle and Isocrates on oratory. Any reader will agree that Cicero's delightful essay bears no resemblance to a handbook.

Despite his distinction as a stylist, Cicero was as much a dilettante as the encyclopedists in the broad scope of his reading and interests. He particularly admired the syncretistic philosophers of the second and first centuries B.C.—Panaetius and Posidonius among his favorites— whose readily digestible doctrines appear to have been written for wide audiences. Cicero knew very little pure science, and many of the scraps of information that he passes off as derived from reading the mathematical works of the masters he undoubtedly got at second or third hand from books of contemporary philosophers. For example, Cicero's friend Atticus, writing from Athens, importuned him to write a book on geography and sent a copy of a geographical work in Greek by Serapion to assist him in his researches. Cicero replied with thanks for the book and promised to produce a geographical treatise, but admitted in confidence (*Ad Atticum* II.4) that he comprehended hardly a thousandth part of the matter in Serapion's book. Another letter (II.6) finds Cicero uncertain about fulfilling his promise to Atticus; a book on geography is a large undertaking and Cicero is disturbed to find that Eratosthenes, his most trusted authority, has been criticized by Serapion and Hipparchus; geography is a difficult and tedious subject, and it does not provide the writer with an opportunity to expatiate in a florid style. In the next letter (II.7) Cicero is still toying with the idea of writing a book on geography, but that is the last we hear of the project. It is probable that Cicero's knowledge about Hipparchus' criticisms of Eratosthenes was derived from reading Serapion and not Hipparchus.

Cicero, like all his compatriots of whom we have intimate knowledge, lacked an understanding of the methodical and exacting nature of scientific research. To the Romans the mathematical geniuses of Greece were awesome freaks, men of remarkable achievements but a rather pitiable way of life. Cicero discloses his personal views, which are quite unsympathetic to the mastery of scientific disciplines, in the early chapters of his treatise *On the Orator* (I.3–6, 11–16), briefly summarized here. He observes that records of the past reveal that many philosophers were experts in some one field and at the same time were able to encompass the entire range of human knowledge. Mathematics, he says, is an obscure field, an abstruse science, complicated and exact; yet so many have attained perfection in it that we might conclude almost anyone [!] who seriously applied himself would achieve a measure of success. The Romans have surpassed all other nations of the world in their natural ability in oratory. Eloquence consists of many components, which Cicero enumerates; but the first on his list is a thorough knowledge of very many subjects. Mere eloquence, without mastery of the subject, will make the speaker appear vapid and ridiculous. Cicero would not go so far as to say that an orator, caught up in the hurly-burly of Rome, should not be able to indulge in the relaxation of ignorance on some subject; but the very name 'orator' seems to imply not only the profession of eloquence but the ability to speak with full knowledge on any subject proposed.

Cicero has one of his speakers assert that an orator who has prepared his briefs carefully will speak on any subject with more skill than an expert or a man who has made an original discovery in that field. The architect Philon, who constructed the arsenal at Athens, himself explained his building plans to the Assembly; but it would be a mistake to suppose that the success of his presentation was attributable more to his skill as an architect than as an orator. If, as men of learning are aware, Aratus was able to compose a polished poem on the heavens, though he was ignorant of astronomy, and Nicander wrote a celebrated poem on farming, though his life was remote from the country, both by reason of their poetic faculties, why, he asks, should it not be admitted that an orator is able to speak with eloquence on subjects which he has studied in preparation for legal cases? The art of the poet closely resembles the art of an orator; neither accepts any limitations to his prerogative of expatiating in any field of knowledge he may choose.

Such views would naturally irritate a modern scientist. Cicero's line of argument draws uncomfortably close to the attitude of a facile lawyer who in handling accident cases has acquired from medical reference books so many bits of anatomical, physiological, and neurological information and has so often succeeded in confuting the professional testimony of doctors in court that he sincerely believes he "knows more medicine than the doctors." It would be wrong, however, to depreciate Cicero and his fellow Romans because they did not grasp scientific concepts. Instead, our admiration for the Greeks should increase. Men like Hippocrates, Archimedes, and Hipparchus should be regarded as prodigies in the early development of scientific knowledge. In Greek society the scientists were more abstruse than the philosophers, who were often engaged in trying to interpret them; and the Romans were merely following the example of Hellenistic intellectuals who were satisfied to embrace the results of Greek scientists without understanding their methods.

The two works of Plato that aroused the greatest interest among Platonists in Greece, the *Timaeus* and the *Republic*, both received special attention from Cicero. He translated the *Timaeus* and imitated the *Republic*, producing a masterpiece of his own by the same name. Cicero's *Republic* is realistic by contrast with Plato's work, bringing to bear considerable personal experience in dealing with Roman politics, whereas Plato was discussing an ideal state. The closing episode of Plato's *Republic*, known as "the myth," or the Vision of Er, became the most celebrated part of the work and was the subject of separate Greek commentaries. In this episode the reader gets a glimpse of the heavens and of life hereafter in the form of a narrative told by a soldier who arose from the dead after lying on the battlefield for eleven days. Plato was censured in later times for resorting to the device of a resurrected informant. Wishing to include a glimpse of the heavens in his own *Republic*, Cicero accomplished it more circumspectly, through a dream related by the Younger Scipio. The *Dream of Scipio* is a majestic piece of prose, reaching a pinnacle of style among his writings. Its cadences and phrases were compared to organ music by J. W. Mackail, who felt that only Virgil attained such nobility of expression in the Latin language.[6] It was published in separate editions after Cicero's death and, like Plato's myth, attracted the attention of commentators. Two commentaries have survived, and from a statement made by Macrobius [7] we assume the existence of others.

It is important to note that Cicero derived his conception of the heavens not from Plato but from Posidonius. He prefers the Posidonian order of the planets, which located Mercury and Venus between the earth and sun, to Plato's order, which placed them above the sun; and he embraces Cratean doctrines, found in Posidonius' works, of the four inhabited quarters of the globe, with an ocean and torrid zone to keep them eternally separated. He also derived from Posidonius the conception of the Roman Empire as a reflection of the commonwealth of God and the heavens as a place of reward for statesmen and philosophers who had conducted their lives nobly on earth. Cicero had shown sporadic interest in the subject of cosmology from the time when, as a young man, he had translated Aratus' *Phaenomena*.

The last figure of the Republican period we shall discuss here is a man who is usually regarded as one of the most brilliant Roman intellects and is frequently singled out for special attention in even the most cursory surveys of ancient science. In Lucretius (c. 99–55 B.C.) we see a genius whose remarkable perspicacity in some matters and ignorance or naiveté in others pose a problem for scholars who would assess the man accurately. The profundities of his poem *On the Nature of the Universe* fill us with almost as much amazement and awe as Lucretius himself experienced in contemplating the world about him. But if we focus only upon the remarkable intuitive and perceptive flashes of his genius, as most historians of science have done, and fail to regard the man in the light of his scientific and philosophical background, we cannot appreciate the anomalous character of his speculations and theories.

The breath-taking sweep of Lucretius' range, from "the flaming barriers of the universe" to subtle and tender observations about the tiny creations of the animal, plant, and mineral worlds; his reverence for the mystic kinships existing in the megacosmic and microcosmic realms, suggest on the one hand the Hebrew poets and prophets and on the other Walt Whitman. His guesses-in-the-dark were fortuitous, and his indebtedness to his predecessors is undoubtedly heavy. Yet Lucretius encompassed in one book a doctrine of separate worlds, each with its own sun, moon, and sky, coming into being and passing out of existence —a view akin to the continuous universe theory of contemporary cosmologists; doctrines of lightning-like motions of atoms, buffeting each other within the confines of a hard body; of the infinity and indestructibility of matter, of the equal velocity of bodies falling in a vacuum; principles, anticipating Darwin, of the struggle for existence and survival of the fittest; and views substantially in agreement with those of present-day anthropologists regarding the modes of living of primitive men and the beginnings of language and the arts. This would mark Lucretius as a notable figure in any history of early science. And to have enhanced his majestic contemplation of the physical universe with the surging of resonant hexameter lines which link in exquisite partnership the wonders of nature with one of the most sonorous and liquid vocabularies ever to pour forth from the human mind has given his work a luster that has made it one of the finest poems in any language. This is the Lucretius who will continue to excite the imaginations of thinkers and poets of the future as he has done in the past. This is the Lucretius that most historians of literature and science have admired. But unfortunately the facets of Lucretius that come to our attention in the present volume are lusterless ones.

It is now known that the best of his scientific observations as well as the worst were not original with him but came from his reading. Lucretius was an ardent adherent of the Epicurean school and, like his master Epicurus about two and a half centuries before him, felt nothing but contempt for philosophers whose views were not in accord with his own. The Epicureans drew most of their physical theories from the atomists Leucippus and Democritus and the speculations of early Ionian philosophers. They shut their eyes to scientific developments that had taken place during the Hellenistic Age. Whereas Stoic and Platonist philosophers had been keeping up to date by incorporating the elements of Hipparchan astronomy and Euclidean geometry in their systematic treatises, the Epicureans stubbornly subscribed to views held by physical speculators who were, for the most part, contemporaries of, or earlier than, Socrates.

Epicurus appears to have realized the deficiencies of his knowledge of astronomical subjects. He relegated celestial phenomena to the class of "unclear" objects because we never get a close view of them. When Epicureans were unable to apprehend objects directly by the senses, they were obliged to postulate theories that did not contradict sense perceptions and to assume that if such theories did not operate in this world they would be operating in some other world. So they evolved a procedure of accounting for phenomena beyond the ken of their senses by offering several alternative explanations. Lucretius suffers from this handicap, and a further one common for a Roman adherent of a Greek system: he fails to comprehend the more abstruse doctrines of Epicurus and has an obvious lack of interest in astronomical matters. The disastrous effect upon his cosmology is seen in the puerile conceptions he sets forth in Book V of *On the Nature of the Universe*.

The world in which Lucretius lives is geocentric. The earth is motionless at the center of a hollow sphere. He is uncertain whether celestial bodies move independently of or together with the rotation of the heavens, and offers five possible forms of extra- or intra-mundane air currents to propel the heavens; or perhaps celestial bodies move spontaneously to seek nourishment. The earth is rounded and flat; it gradually thins out on the underside to coalesce with the air beneath, which supports it. The sun's orb and heat are not much greater than they seem to us to be; the moon is also the size that it appears to be—a breadth measured in inches. (Several writers who made little claim to being authorities in scientific matters ridiculed Lucretius for his statements about the size of the sun and moon.) It is possible that the sun is driven in its course from Cancer to Capricorn and back by shifting currents of air. The moon and stars may be similarly propelled. Darkness may be caused on earth by the sun (1) being extinguished at the end of its day's journey or (2) being carried beneath the earth during the night by the same force that moves it overhead during the day. It is possible that a new sun may be formed in the east every day by fires gathering to kindle it. The inequality of length of days and nights in spring and fall may be accounted for by supposing (1) that unequal arcs are described by the sun in passing above and below the earth at different seasons; (2) that the air is thicker in certain regions of the sky, so that the sun's light cannot penetrate so rapidly at some seasons as at others; or (3) by the difference in amount of time required by the fires to kindle a new sun in the east. The moon's phases may be explained by hypothesizing (1) that its light is borrowed and that it changes position with respect to the earth and sun, exposing varying amounts of its illumination to us; (2) that the moon shines by its own light, which it reveals in various forms; (3) that some opaque body accompanies the moon and obscures it in varying degrees; (4) that the moon is by nature half illuminated and turns the bright side gradually toward or away from us; or (5) that a new moon is created every day. Solar and lunar eclipses may be explained (1) by the interposition of the earth or moon as obscuring bodies; (2) by some other opaque body being interposed; or (3) by the illuminated body passing temporarily through regions that extinguish its flames or brilliance.

Scholars have suggested that Lucretius may have derived some of the correct alternative views—which he usually places before the startlingly naive conceptions in each group—from Hellenistic handbooks or other works written after Epicurus' time; but few of his astronomical notions could not have been found among the doctrines of the early Ionian philosophers. The anomalous and derivative character of the scientific portions of Lucretius' poem makes it reasonable to conclude that his significance should be judged as a poet, not as a scientist.

VI

EXPANDING HORIZONS IN THE AUGUSTAN AGE

As soon as Octavian, by his victory at Actium in 31 B.C., had decisively crushed the ambitions of Anthony and Cleopatra to become co-rulers of the Roman Empire, he was prepared to undertake with full concentration and vigor a program of expansion and consolidation of imperial boundaries and of administrative and public reforms that had been planned and inaugurated by Julius Caesar two decades earlier. By the time Octavian assumed his new title of Augustus, a few years later, Roman citizens throughout the empire were coming to realize that a new age had been ushered in, that a sense of direction was being given to their affairs, and that they themselves had clear responsibilities to meet.

Even the poets, often against their will, were caught up in the new spirit of the times. Virgil had been invited to become an official poet of the new regime and, happily situated with a liberal pension and commodious villa provided by the emperor, was devoting his efforts to composing an epic on the new order. Like Lucretius a generation before him, he had had a thorough training in Epicureanism, but he could not for a moment have entertained such an attitude of indifference to temporal authority as was advocated by Epicurus and maintained by Lucretius. The love poets, too, responded to the changing times. During the decadent years of the dying republic, Catullus and his madcap coterie of versifiers had reveled in the excesses of the fashionable set and the perversities of the demimonde. Occasionally one of them, out of a desire to get away for a time from the giddy whirl at Rome, accepted an invitation from a government official to join his entourage on its way to some distant province. Love poets during Augustus' reign were no less torrid in their love affairs, but in melancholy moods they would think of the stern call to duty that might come at any moment. In the midst of drinking bouts they could impress their sweethearts by drawing diagrams of battle formations in wine spilled on a tavern table or by making hazy sketches of faraway places to which they were being sent on military assignments. The new bêtes noires of an Augustan elegist's life were the long overseas stay and the barbarian javelin that might be destined for him.

Augustus appreciated, as had Julius Caesar, that good communications were essential for a strong empire. Highways were being laid throughout the provinces, and Augustus was learning that shortages in military manpower could be offset by being able to move crack troops to vulnerable frontiers within a few days or weeks. He established a rapid courier service, locating horse and carriage stations at regular intervals along main arteries. Messengers and officials traveling on imperial missions could average fifty miles a day in emergencies. To use his vast network of highways to best advantage, Augustus needed detailed maps: so he reconstituted a project, outlined and begun by Caesar, of mapping the entire known world, and assigned the task to one of his leading ministers, Marcus Agrippa. Agrippa instituted an elaborate imperial survey, computing mileages from the figures marked on milestones along the highways, to determine the dimensions of each province and the total distances across the empire. This ambitious undertaking required more than twenty years to complete. Five years after Agrippa's death a large master map, incorporating the data of the survey, was affixed to a wall of the Portico of Vipsania in Rome; it became the archetype of copies that were distributed freely to ranking military and administrative officials. Agrippa's survey collated vital information of all sorts, which was then deposited in government archives and used in the compilation of commentaries to accompany the maps. One late and faulty copy of the map has survived, a parchment roll twenty-one feet long and a foot wide called the Peutinger Table. It covered the known world from Britain to China. The greater part of Britain and Spain is now missing as a result of the disappearance of the first segment.

Agrippa's survey could have provided Roman geographers with the raw data for accurate regional accounts and for reasonably good representations of the known world. Pliny the Elder had access to the maps and data; but although he appreciated the care with which the survey had been conducted and frequently adopted its figures for his distances, he did not begin to realize its value as a basis for his geographical books. Like other Roman geographers he was more interested in readability than in reliability; instead of producing an up-to-date geography of the Roman Empire and of the exterior world which was becoming known through commercial and diplomatic contacts, he reverted to the outmoded style of periplus manual originated four centuries earlier by the Greeks: he conducts readers on a survey of coastal regions, and ransacks Greek writers who since Herodotus' time had been giving credence to tall tales about strange peoples and lands.

Augustus' military campaigns carried him into interior regions of Europe, Asia, and Africa, in the subjugation of Alpine peoples, the consolidation of frontiers along the Rhine and Danube rivers, and penetrations into Arabia Felix and Ethiopia. We would naturally expect Roman geographers to give special consideration to newly conquered regions that were making heavy demands on manpower and resources and were engaging most of the attention of provincial administrators and generals. Yet Pliny, though he served in military campaigns in Germany and wrote a history of the German wars in twenty books, devotes less than two pages of the geographical portion of his encyclopedia to that country and presents an almost useless summary of information—a consummate example of hidebound adherence to handbook traditions. The amount of attention he gives to a country is largely determined by the interest it held for Greek readers of the Hellenistic Age.

The Greeks, in direct contrast to the Romans, had always been a maritime people, who kept in touch with nearby Aegean city-states, and with distant colonies that dotted the shores of the Mediterranean and Black seas, by means of trading vessels that skirted the coasts or crossed seas, where closely set islands would afford them shelter during storms. Greek geographical writings reflected their maritime instincts and practical needs. Evincing little or no interest in regions in the interior, early Greek geographers ranged along coastal countries, taking them up in the order a navigator would come upon them in making a circuit of the Mediterranean or eastern seas. Reports of sailing expeditions along the Atlantic coasts of Europe and Africa and along the shores of the Red Sea, Persian Gulf, and Indian Ocean were published separately and were digested and incorporated into later handbooks. There was, for instance, the *Periplus* of Scylax of Caryanda, an account (c. 350 B.C.) of a voyage starting at Cadiz, continuing along the coasts and bays of the northern Mediterranean, around the Black Sea, down the Levantine coast, across the North African coast, through the Straits of Gibraltar, and along the West African coast to the island of Cerne, off Rio de Oro—a sort of coastal pilot's guide. In its present form this account is the product of a later hand, a recasting of what was once a composite of the reports of a number of voyages.

Such periploi had much to do with determining the form and arrangement of Greek geographical treatises. Lay handbook authors adopted the periplus framework for dealing with the main subject, which was regional geography. They also regularly opened their discussion with a consideration of topics of geodetic interest—the earth's shape, size, and position, the celestial circles, variation of climates and seasons, and locations and extent of inhabited regions—a practice that was instituted by some mathematical geographer, perhaps Eratosthenes or Eudoxus. And probably because of the fascination that the reports of strange peoples and customs by Herodotus, Ctesias, and Megasthenes had long held for Greek readers, and because of the interesting anthropological digressions in the writings of Polybius and Posidonius, accounts of weird customs and astonishing phenomena of faraway people and places also came to be regular features of the geographical handbooks. Homer probably inspired this fad, as he did so many others in geography, by his mention of pygmies and Ethiopians. Surprisingly enough, Greece itself was given little attention by Greek geographers; it was considered too familiar for detailed treatment.

With the greatly expanded geographical horizons resulting from such expeditions as the voyage of Pytheas around Britain into the North Sea (c. 300 B.C.) and from Roman conquests of large continental areas, geographical authors who knew only the coastal survey method of handling regional geography were hard pressed to accommodate this meager framework to the new world geography. An absurd conception arose of the known world as two great continental land masses, virtually bisected by the Mediterranean, the whole surrounded by one ocean. The coastal survey could continue to proceed with meticulous care around the bays and inlets of the Mediterranean; but in eastern regions mountain ranges like the Caucasus were surmounted with scarcely a mention; and from India the reader was swept with breath-taking rapidity across the Northern Ocean to find himself at the Straits of Gibraltar. Similarly, the survey of the southern half of the known world was conducted painstakingly along Mediterranean shores, indistinctly in the eastern world, and ended up with a pell-mell sweep in a sentence or two across the shores of the southern ocean, from Ethiopia to Mauretania.

To demonstrate to readers that the northern and southern halves of the known world were surrounded by water, and thus to make the periplus appear to be a reasonably adequate scheme for handling the countries of that world, handbook authors introduced tales of fantastic voyages, presented as if they were sober truth—such as the report of Indian sailors blown off their course past the Caspian to the shores of Germany, and of the circumnavigations of Africa by Hanno and by the mariner (not the mathematician) Eudoxus. Such periploi ignored virtually all of central Europe and the great expanses of interior Asia on the way to India—countries like Media, Parthia, Bactria, Margiana, Sogdiana, and Drangiana, described by earlier Greek writers who were drawing upon the historians of Alexander's expedition and were not restricted by the periplus method.

Since geography was mainly a practical subject, it never became an absorbing field of interest to Greek scientists; their geographical treatises do not measure up to the standards of their other scientific writings —but they did respond to the interests of Greek readers. The Romans, who did not take naturally to navigation and who had a vast empire to maintain in the interior of three continents, still followed the maritime form of Greek geographical treatises. Since their grasp of mathematical geography was weak, their mathematical introductions were usually meager and faulty. And among their borrowings from the anthropological digressions of Greek historians and geographers, the Romans preferred the most incredible and miraculous stories.

The *Chorography* (commonly referred to as *De situ orbis*) of Pomponius Mela is the earliest extant Latin treatise on geography. A compressed work, only eighty pages in length, it exhibits most of the faults of a Latin handbook in this field. The author remarks on the tedium of a subject that necessarily involves the use of many place names, and makes obvious efforts to write divertingly. A reference to the subjugation of Britain, which was then in progress, dates Mela's work about A.D. 43, a time when Roman dominion extended from the southern part of Britain to Armenia and Ethiopia; yet with a few exceptions most of his information could have been found in the three books of Eratosthenes' *Geographica*, written nearly three centuries earlier.

It has been surmised that Mela was interested in Spain because he was born there, or because he depended upon a work of Varro, who had been one of Pompey's lieutenants warring against Caesar in Spain.[1] It is more probable that for the most part Mela, like Varro and Pliny, was merely following Greek models. Spain had long been a region of special interest to the Greeks. In the sixth century B.C. Massilian Greeks, striking westward, had established colonies on the east coast of Spain. After Roman annexation of the Carthaginian possessions in Spain had brought the entire Mediterranean area into close contact, Greeks like Polybius and Posidonius were able to travel extensively in the peninsula and to record their observations for curious readers at home.

On the other hand areas of primary military importance to the Romans—the interior of Germany, the Alpine and Danubian provinces, and Dacia—get no mention by Mela; and other interior regions receive very scanty attention. He observes that German geographical names are too barbarous for Roman lips to pronounce.[2] Athens, he says, is too famous to require attention. He finds space however for garbled mention of fabulous pygmies, winged serpents, the phoenix, the Amazons, Hyperboreans, headless Blemmyes, and goat-footed Aegipanes; some of these mythical creatures had been described by Herodotus, and Mela may have been introduced to them by that venerable authority.

Mela displays no interest in or knowledge of mathematical geography; but because it was customary to introduce geographical treatises with a discussion of the earth's shape, zones, and habitations, he gives token attention to these topics at the opening of Book I. The earth is a globe, situated at the center of the universe. The known world is twice as long as it is wide. No dimensions or estimates of distances are given, although these are a standard feature of the Greek handbooks. The earth has five zones, of which the northern and southern temperate zones are inhabited. Mela does not take the trouble to give conventional treatment to the earth's four habitations, two in the northern and two in the southern hemispheres; instead, he refers vaguely to antichthones, inhabitants of the southern temperate zone, who are ever separated from the northern world by the intense heat of the torrid zone. The known world lies in the northern hemisphere, surrounded by continuous ocean. Opening into the ocean are four large seas, the Caspian on the north, the Persian and Arabian gulfs on the south, and the Mediterranean on the west. *Terra cognita* is divided into three continents, Europe, Asia, and Africa, with the Tanais (Don) and Nile rivers serving as boundaries. All of Mela's introductory material, with the exception of his reference to inhabitants in the southern hemisphere, had been a familiar part of the handbook tradition since Eratosthenes; the doctrines about habitations in other parts of the globe, originated by Crates in the second century B.C., shortly became a regular feature too.

Immediately following the brief introductory account of the earth as a whole and of the seas, continents, and countries of the known portion, Mela begins his coastal survey of the Mediterranean. The importance he attaches to this part of his work is indicated by his allotting two-thirds of his entire treatise to it. He starts at the Straits of Gibraltar, takes up the countries along the North African coast from Mauretania to Egypt, crosses to Arabia and passes along the eastern Mediterranean, then enters the Black Sea, which he regards as part of the Mediterranean, and makes a circuit of it. Adherence to a periplus scheme obliges him to deal with Spain and Gaul a second time, when he comes to discuss the countries lying along the Atlantic and northern oceans. There follows a separate section on the islands of the Mediterranean, including some mere rocks in coastal waters. Mela's ignorance of interior regions is thorough. He repeats the age-old misconception of the Danube dividing into two branches, one of them emptying into the Adriatic; and he supposes that the Alps bound Germany. His detailed description of the facing rocks at the Straits of Gibraltar and knowledge of the region around the cape may indeed be reasonably attributed to his having been born in the vicinity.[3]

At the start of Book III Mela is ready to range out along the Atlantic coast. This brings him to a discussion of the tides, a subject that became customary in the handbooks after Posidonius. He is baffled by tidal phenomena, describes them and the effects badly, and ends by suggesting explanations that faintly echo the collision theory of Crates and Posidonius' theory of lunar influence. Mela's survey of the coast of western and northern Europe is the first one extant that represents the Iberian peninsula as jutting out in the northwest and forming, with its great promontory, the distinct indentation of the Bay of Biscay. Strabo had been unaware of this projection and had conceived of the Atlantic coasts of Spain and Gaul as extending in an almost straight line. Mela also has a good idea of the shapes and relative positions of Britain and Ireland and makes the first known mention of the Orcades (Orkney Islands). His treatment of Thule is the conventional one, drawing attention to the short nights in summer and absence of darkness at the solstice; he is confused in supposing that there are short dark nights in winter. His book is the first we know to have described the Codanus Sinus, a large bay east of the Danish peninsula, containing many islands, among them that of Codanovia, which many have identified with Scandinavia.

Being the first author known to describe or name certain places in northern Europe does not necessarily reflect credit upon Mela as a geographer; it merely happens that the sources that he was using for those regions have not survived. He also had precise information about localities along the Caspian Sea and yet, as we have noted, he regards it as a gulf opening upon the northern ocean. Mela cites no less an authority than Hipparchus for the plausibility of conceiving of Taprobane (Ceylon) as the beginning of a continent in the southern hemisphere. It would be rash, however, on the basis of this reference, to attribute this archaic conception, discredited after Alexander's expedition, to a geographer of the caliber of Hipparchus; he was frequently cited, because of his reputation, by Latin compilers who had only vague notions of the contents of his works. Because Cornelius Nepos is twice cited by Mela, some have regarded Nepos as a direct source. This is possible; but there is the coincidence that Pliny too cites Nepos in the same connections. It is reasonable to suppose that such tags of information were identified with certain authorities and became part of a common stock of handbook materials. In the passages in which Nepos is cited by Mela and Pliny, the early Carthaginian navigator Hanno is also referred to as an authority. We have here another of the many bits of evidence that Pliny and Mela were using works ultimately derived from the same source.

Before turning to Pliny, whose discussion of mathematical and regional geography resembles Mela's treatment very closely in spite of being much more elaborate and detailed, it is well to review three authors whose works belong to the handbook tradition and throw considerable light on it: Vitruvius on architecture, Celsus on medicine, and Seneca on physical science. Although Seneca is a generation later than the Augustan Age, it is convenient to consider him here with others whose works need not be described in detail.

Vitruvius, according to his testimony in *On Architecture*, was a practicing architect and government overseer of military engines. His work, published probably between 26 and 14 B.C., is the only treatise in this field to survive from antiquity. It is a diffuse compilation in ten books, which takes a very broad attitude about the subjects that belong in a work on architecture. Its uniqueness and apparent authoritativeness made it one of the most intensely studied and influential of classical works in Renaissance times. Opinions about Vitruvius' competence as an architect and about the merits of his book differ sharply. Some think that instead of being an architect of the Augustan Age, as he represents himself, the author of *On Architecture* was a sort of "pseudo-Vitruvius" of around A.D. 400;[4] a mere compiler with no critical ability to assimilate or interpret his sources and no literary ability to present his materials. Others call the volume a remarkable manual by an accurate observer of the architectural and building details of edifices familiar to him.[5]

There is basis in one part or another of this anomalous book for such widely divergent opinions, although a date of A.D. 400 is admittedly an extravagant conjecture. Some of Vitruvius' admirers among professional architects and historians of architecture have nevertheless been aware of the embarrassing blunders that Renaissance neoclassical architects made because they could not comprehend his badly written accounts or were too respectful of the authority of a man who was himself confused and unable to interpret his Greek precursors.

Vitruvius' work dwells more upon Hellenistic architecture than upon Roman, a natural circumstance if it was derived mainly from Greek sources and was compiled before Augustus' extensive building program got under way. Correspondences have been found in the treatises of Vitruvius and the mechanician Heron of Alexandria which indicate that both were drawing upon the same [ultimate] Greek sources. It is a reasonable assumption that all the theoretical portions of Vitruvius' book came from Greek authors. There is no way of ascertaining whether Vitruvius depended heavily upon Varro or other Latin writers as intermediaries or went directly to Greek materials. Book IX of Varro's encyclopedia may have been a direct source. Vitruvius includes Varro's name at the end of a long list of authorities, nearly all of whom are Greek; but from what is now known about Latin compilers' habits, citation of an authority is not dependable evidence that he was ever consulted. The treatise exhibits the familiar defects of a Roman compilation on a theoretical subject. The author is fervent in arguing the values of theoretical training for architects but is himself vague and incompetent in presenting Greek theory. Many topics are introduced, but most are handled inadequately. A notable exception is the chapter on the principles of sundials (IX.7), a summary account which, because of its technical expertness and its interposition in the mélange of unrelated subjects that constitute Vitruvius' book, must be assumed to have been culled from some Greek manual.

The opening chapter discusses the proper education of an architect and recommends a Greek type of training in the liberal arts, embracing study of grammar, geometry, history, philosophy, music, medicine, law, astronomy, optics, arithmetic, painting, and sculpture. Vitruvius is able to offer cogent arguments on the importance for an architect of some studies, such as geometry, but about others he is confused and not very convincing. To affirm the importance of a knowledge of the theory of music, he points out that military engineers determine the proper tension of the sinews of catapults by listening to the pitch of the vibrating cords. And he observes that architects of Greek theaters placed bronze vessels, arranged according to the harmonious chords of the fourth, fifth, octave, and double octave, beneath the seats in order to give greater resonance to the voices of actors. This worthless expedient, incidentally, was tried in the choir lofts of many medieval churches.[6] In spite of his advocacy of geometrical studies in an architect's training, Vitruvius gives discrepant approximations of 3 and $3\frac{1}{8}$ for π. The Archimedean approximation of $3\frac{1}{7}$ was the accepted figure, and Vitruvius could have checked his value by measuring one of the many columns he describes in his work.[7] He recommends a knowledge of law for its utility in drafting building contracts, of astronomy for the orientation of buildings, and of arithmetic for computing construction costs. Eminently practical and woefully impractical, at times meticulous and at other times uncritical and careless, Vitruvius shows some of the worst features of a Roman compiler together with some highly commendable characteristics of a practicing architect.

A separate section on astronomy opens Book IX, but instead of focusing upon aspects related to the orientation of buildings, Vitruvius inserts a conventional handbook treatment, so condensed and deficient in details and figures as to be of little value to an architect. The heavens revolve about the earth on a celestial axis, the north celestial pole being far above our heads, the south pole beneath the horizon. The broad belt of the zodiac passes slantwise (the degree of inclination not given) across the middle of the heavens, six of its twelve signs always above and six below the horizon. If a portion of a sign is emerging on the eastern horizon, an equally large portion of another sign is buried beneath the western horizon. The periods of planetary revolutions are noted, and here Vitruvius states in figurative and ambiguous language that "the paths of Venus and Mercury are wreathed about the sun's rays as their center," which, he says, accounts for the peculiar retardations, stations, and retrogradations of those planets and for their shifting back and forth with respect to the sun. This description of the behavior of Mercury and Venus has often been interpreted as an unequivocal statement of Heraclides' theory of the heliocentric motions of those planets. Heraclides' theory crops up in a number of places in the handbook tradition, usually in a form that does not correctly represent it; and Vitruvius' statement does indeed reflect it faintly, but with an admixture of other doctrine, as we shall see.

Discussion of the periods of planetary revolutions is resumed with a brief account of the orbits of Mars, Jupiter, and Saturn—which calls to Vitruvius' mind the peculiar solar anomaly in their motions. During the second century B.C. Greek astronomers had become aware that the retrogradations of the superior planets occur when they are on the opposite side of the earth from the sun and that the sun thus has a regulating influence upon the motions of all planets (except the moon in a geocentric orientation). This awareness did not lead them to reaffirm the correctness of Aristarchus' hypothesis of planetary motions, although that would have eliminated the need to account for retrogradations and stations. Astronomers after Aristarchus were more interested in the mathematical aspects of their subject than in physical realities. But when Stoic philosophers interested themselves in astronomical matters and began introducing the theories of Hellenistic astronomers in their handbooks, they associated the solar anomaly of planetary motions with Stoic doctrines.

Cleanthes, successor to Zeno, the founder of the Stoic school, had propounded with religious fervor the doctrines that the universe was a living, sentient being, of which the sun was the animating principle, the heart and ruling force, and the abode of God. According to a common belief of Stoic astronomers, the cold, immobile earth was at the exact center of the *corporeal* universe, but the sun, with its magnitude and intense heat, was the ruling force of the *living* universe, exercising a controlling influence on the motions of the planets. It is clear that Posidonius embraced these Stoic doctrines, and that probably explains why they are found in the writings of Cicero and Plutarch, and in the handbooks of Pliny, Seneca, Theon, Chalcidius, and Macrobius. Vitruvius calls attention to the fact that the retardations of the superior planets begin when they enter the fifth sign away from the sun and, like Pliny, he stresses the powerful effects of the sun's rays as the cause of the stations and retrogradations. His statement about the paths of Mercury and Venus being "wreathed about the sun's *rays* as their center" makes it plain that he is not clearly expounding Heraclides' theory but is confusing it with Stoic doctrine.

Vitruvius does not give precise details about the planets, such as the inclination of their orbits to the ecliptic or the maximum elongations of Venus and Mercury. In a chapter on the moon he discusses only its phases, offering two popular explanations. First he states the theory of Berossos the Chaldaean, that the moon is a half-illuminated orb and that the attractive power of the sun's rays upon the illuminated side, a sympathy of light for light, is responsible for the phases. The second, and correct, explanation he attributes to Aristarchus of Samos, though it was offered by some pre-Socratics and was widely accepted before Aristarchus' time. Another chapter traces the sun's course through the signs of the zodiac, marking the solstitial and equinoctial points as the sun has invaded the eighth degree of the signs of Aries, Cancer, Libra, and Capricorn. (This use of the eighth degree is believed to have been introduced into Greece from a Babylonian system about the time of Hipparchus; before that time the fifteenth degree, set by Eudoxus, had been used.)[8] Vitruvius omits many of the standard features of handbook astronomy, such as the anomalies in the four quadrants of the sun's (apparent) orbit, or the supposed inclination of the orbit to the ecliptic, or the effects of the sun's course upon the light and climate of the habitations of the globe. He closes his astronomical section with a drastically condensed description of the constellations of the northern and southern hemispheres. Marked resemblances to the arrangement and details of Aratus' *Phaenomena* indicate that Vitruvius was following some conventional handbook in the Aratean tradition. An ancient reader would have had to depend heavily upon the maps that accompanied Vitruvius' text to be able to relate his descriptions to the heavens.

Cornelius Celsus, the next author to be considered in this period, was the third eminent encyclopedist to appear in the Latin world of letters. He had Cato's, and not Varro's, attitude about the subjects that were fit to engage the attention of a Roman encyclopedist. Celsus' work covered six *artes*: agriculture, medicine, the military arts, oratory, philosophy, and jurisprudence—all subjects, philosophy excepted, that were handled by Cato in separate manuals. However, the eighth book in Varro's *Nine Books of Disciplines*, that on medicine, may have been one of the sources, though hardly an important one, for the medical portion of Celsus' large encyclopedia. Varro took a bold step when he introduced to the Roman reading public an encyclopedia prepared in the Greek style, adding to the seven purely liberal arts of the Greeks two practical arts, medicine and architecture. It should be pointed out, however, that the subjects added by Varro, though practical by comparison with the *artes liberales*, were still distinctly Greek *technai*, fields which the Romans themselves always identified as Greek professions and not characteristically Roman artes, like military tactics and agriculture. The facts that *On Medicine* is the only one of Celsus' six artes to survive and that manuals on medicine of one sort or another figured prominently in the Latin handbook tradition into the Middle Ages confirm the belief that a knowledge of medicine was considered an important part of the store of general information which ought to be familiar or available to any responsible and provident Roman *pater familias*.

The highly disputable features of Celsus' manual, like those of Vitruvius' work, furnish scholars with an abundant store of ammunition to fire at each other while maintaining their readily defensible positions of interpretation. The chief enigma about Celsus, which underlies the controversies, is that the book is a masterpiece of Latin prose style and, in arrangement and treatment of subjects, attests to a writer of superior intellectual capacities, not to a pedestrian compiler translating or adapting a Greek original.

Celsus' style has evoked the praises of Latinists of every generation since the rediscovery of his work in the Renaissance. Humanists at that time, with their keen appreciation of pure Latinity, compared Celsus to Cicero. The first printed edition of his work appeared in 1478, before the works of Hippocrates and Galen. Since that time he has been ranked as the third most important medical writer of antiquity, after those two great masters. When the Swiss alchemist and physician Theophrastus Bombastus von Hohenheim was seeking a distinctive Latin appellation, he called himself Paracelsus, "more eminent than Celsus." Today Celsus' work is regarded, on the basis of its form and content, as the most satisfactory book in the entire collection of extant Latin scientific writings and, by some critics, as the most orderly and best arranged single treatise in the entire field of ancient scientific literature, Latin or Greek.[9] Because *On Medicine* covers the field so neatly and efficiently, without becoming involved in the controversies of dogmatic Greek sectarian practitioners, some scholars have been inclined to emphasize the Roman traits they find in the work; and recently some highly competent medical historians have argued that Celsus was a physician.

The preponderance of scholarly opinion, however, inclines to the opposite view—that he was not a practicing physician. Two German scholars who made a thorough study of his treatise a generation ago, though they differ sharply in their findings, both believe that Celsus' work was substantially a translation or adaptation of a Greek manual: by Menecrates (or Cassius), according to Wellman;[10] by Titus Aufidius Siculus, according to Marx,[11] the most recent editor of Celsus' Latin text. Temkin[12] has since offered strenuous objections to the view that the work is a mere translation; he cites Celsus' ability to select his materials and to synthesize them into a harmonious whole. *On Medicine* exhibits many Latin characteristics, for a book which is substantially a translation or adaptation of a single Greek work; and Quintilian's remark that Celsus was a "man of mediocre intellect"[13] is a slim reed to carry the whole argument that Celsus was merely performing the functions of an editor or translator. His form and style both belie Quintilian's judgment. Furthermore, Quintilian's criticisms of writers generally represent the point of view of a professional rhetorician and often seem unfair to us. Again, the judgment may have been based mainly upon the lost book on oratory, a field in which Quintilian was acknowledged to be the greatest Roman authority.

The evidence seems to point to Celsus having been a highly literate and adept writer, who derived his treatise mainly from Greek sources. The possibility that it was substantially a translation or adaptation of a single Greek work has not been completely ruled out. Celsus freely acknowledges his heavy indebtedness to Hippocrates and Asclepiades, but that cannot be taken as proof that he read their works. The encyclopedia of which *On Medicine* was a part covered a wide range of subjects, and Celsus was regarded in ancient times as a competent authority in many fields. These facts in themselves indicate that he was probably not the primary author of the excellent and highly professional treatise on medicine that bears his name.

We come now to the Stoic philosopher Seneca, tutor and court adviser of the emperor Nero, known to us chiefly for his moralistic writings and his Latin adaptations of Greek tragedies. It is not surprising that a philosopher of the school which subscribed to the doctrine of the all-pervading influence of the Godhead should at times have been drawn to reflect upon natural phenomena. However, his work on natural philosophy betrays the fact that the field was not of major interest to him. *Natural Questions* is a collection of stock information gathered from earlier works, interspersed with moralistic reflections, a few personal observations of nature, and occasional expressions of independent judgment. The topics are haphazardly arranged. There are about a dozen moralizing digressions, ranging from three to eight pages in length, some closely, others loosely connected with the subject under discussion. The strong moral tone of the presentation undoubtedly had much to do with making the book the supreme authority on meteorology and earth science in the Latin West during the Middle Ages. Dante refers to him as "Seneca morale." He was regarded as a Christian by medieval Churchmen.

About A.D. 42, while suffering the heartaches of protracted exile on the desolate island of Corsica, Seneca had turned to natural philosophy to divert his mind from his depressing circumstances. He writes, at the close of an essay addressed to his mother (*Ad Helviam* 20), that he is finding solace in studying the problems of earth, sea, the heavens, and the universe, the very topics of his later scientific work. Twenty years later, when he fell out of favor with Nero and decided to retire from the court and politics, he resumed these earlier interests. Some scholars have believed that he had been collecting materials over the intervening period; but the book itself gives the impression rather of a helter-skelter collection of gleanings from his readings at a time when his mind was disturbed and he was seeking an escape from the world about him. It is likely that the book was written during 62 and 63; there are references in it to events of these years. This was the time in Nero's reign when, according to Pliny the Younger, it was dangerous for a writer to display an independent spirit, and when such topics as the elder Pliny's treatise on acceptable usages of speech and writing were the safest subjects.[14]

Seneca's *Natural Questions* treats in seven books most of the particulars and phenomena of physical geography and meteorology that are dealt with by Aristotle in the four books of his *Meteorology*. Some of the important topics of Aristotle's work are omitted, others are given scanty treatment, still others get disproportionate emphasis and are discussed in several places. Seneca is not nearly so systematic as Aristotle. He cites Aristotle more frequently than any other authority; but the prevailing view at present is that Posidonius, his second most frequently cited source, was his chief authority. Posidonius exerted a weighty influence upon him, directly or indirectly, and is believed to have been largely responsible for the distinctly Platonic characteristics in Seneca's Stoicism. Seneca also refers frequently to a work on meteorology by Asclepiodotus, a pupil of Posidonius. Whether he drew more extensively from this book than from Posidonius will of course always remain a moot question, since both these earlier works on meteorology have been lost. Still another favorite authority of Seneca's, Theophrastus, wrote on meteorology. The original book has since been lost, but it had been translated into Syriac and an anonymous Arab translated it into Arabic and made an abstract of his translation. The abstract, about four pages in length, is now available in German and English translations; it reveals that Theophrastus' work followed the conventional patterns and repeated or reported many of the theories, explanations, and examples that were to become commonplaces in later literature on meteorology.[15]

The relative importance to Seneca, in compiling his work, of Aristotle, Theophrastus, Posidonius, Asclepiodotus, and other lesser writers he

cites will continue to be a subject of speculation. Some of the more san-
guine *Quellenforscher* working in this field have been inclined to believe
it is possible to recast Posidonius' lost *Meteorologica* from the substantial
borrowings incorporated in the writings of Lucretius, Seneca, and Pliny
the Elder.[16] Posidonius is commonly regarded as the chief source of
the meteorological materials of all three. Comparison of their meteorological discussions reveals more differences than similarities, but in a
number of topics—notably lightning and thunderbolts, earthquakes,
winds and their causes and identifications, comets, meteors, and rain-
bows—there are many striking correspondences. That Seneca, who is
fond of quoting Virgil and Ovid, quotes Lucretius only once, although
Book VI of Lucretius' poem is largely devoted to the same topics as
Seneca's work, may be an indication of a dislike for his poem.

Seneca's *Natural Questions* enjoyed its greatest vogue during the
Middle Ages. Together with Pliny's *Natural History*, it continued to be
available to scholars during the Dark Ages and, unlike other classical
works on science, did not have to be reintroduced into Europe from
Arabic sources or discovered by Renaissance humanists in some library
or monastery collection. The highly moral tone and the ethical observations about natural phenomena that endeared Seneca to Christian
writers detracted from the value of his book as a scientific treatise; at
the same time his philosophical reflections on the wonders of nature and
his perception of the comparative smallness of the earth stimulated the
imaginations of Renaissance men like Columbus and Francis Bacon.

VII

PLINY'S THEORETICAL SCIENCE

In whatever connection it is regarded, the vast encyclopedia of natural science by Pliny the Elder (A.D. 23/24–79) is one of the most remarkable of the legacies preserved from antiquity. The usual fate of voluminous—and popular—works has been to become fragmented or to undergo abridgments and epitomes of abridgments; in later ages when scribes did not have the means or fortitude for long copying nor readers the stamina or motivation for the reading, the original work would disappear. But whereas nearly all of Varro's writings have vanished, and only scant fragments of some of Pliny's other works exist, the thirty-seven books of his *Natural History* have survived intact, along with some of its various abridgments.

This was one of the few, if not the only, work of Roman antiquity that deserved comparison as a repository of learning with the prodigious collection of Varro's writings. So far as we know, Varro was the most learned man in the Roman world, both in fact and reputation. Judged by bulk alone, Pliny's 7 works in 102 books are unimpressive compared with Varro's 74 works in 620 books. Varro's attitude toward learning in general was more academic than Pliny's, and he was much more genuinely interested in theoretical science. Our knowledge of what Roman theoretical science there was would be appreciably better if it had been Varro's writings on science that had survived instead of Pliny's. But difficult as it is at present to attempt a reconstruction of Varro's encyclopedia of the theoretical sciences from existing digests and borrowings in later writings, it would be frustrating indeed to undertake to reconstruct Varro's scientific knowledge without the material assistance of Pliny. The *Natural History* will remain the key work in any comprehensive study of what constituted Roman science.

Fortunately for us Pliny was a man both ingenuous and of high integrity, who made no secret of his methods of research and compilation. The self-revelations in his work, together with the information his nephew gives in a letter that intimately records Pliny's scholarly practices and accomplishments, furnish a detailed picture of the study and writing habits of a conscientious Roman encyclopedist. Unquestioning reverence for his uncle, one notes with amusement, did not prevent Pliny the Younger from at least equaling him in candor; and his letter serves to substantiate reasonable inferences about the elder Pliny's methods drawn from an examination of the encyclopedia.

The letter (III.5) opens with a list, in chronological order, of Pliny's writings: a handbook on throwing the javelin from horseback; two books on the life of Pomponius Secundus, Pliny's mentor in his earlier years; twenty books on the wars between the Romans and the Germans; three on the training of an orator, beginning in infancy; eight on grammar and usage; thirty-one on Roman history covering the period of his own manhood; and thirty-seven on natural science. Pliny the Younger expresses amazement that his uncle could accomplish so much, considering that he had a busy law practice, held many high positions in government, was the intimate of emperors, and died at the age of fifty-six. He attributes the achievement to quick insight, incredible industry, and remarkable ability to do without sleep. Pliny used to arise at 1 or 2 A.M., sometimes at midnight, to begin his scholarly researches. If need be, he could drop off to sleep in a moment and at once resume his studies upon awakening. He began his round of official duties before dawn, making his call upon the emperor Vespasian, who also made a practice of working at night. By noon these official duties were completed and he was back at home. After a light lunch and a brief nap, or perhaps a rest in the sun, while making notes or extracts from a book that was read to him, he took a cold plunge bath, perhaps another nap or snack, and then was ready to return to his studies in earnest. No book passed through his hands without his excerpting from it; he maintained that none was so bad it did not have something of value. He worked until dinner; even during the meal a book was read to him and an amanuensis marked passages for copying. Once when a dinner guest chided the secretary for mispronouncing a word, Pliny rebuked the guest for losing ten lines by the interruption. During stays at his country villa Pliny devoted all his time to study. Only the time he was actually immersed in the bath was lost to scholarship, for while he was being rubbed down and dried a book was read to him or he dictated notes. On trips in winter his secretary was by his side in the sedan chair, wearing gloves so that the cold weather would not interfere with reading and note taking. In concluding his account Pliny recalls his uncle once rebuking him for using valuable time in walking when he could have been conveyed and have spent his time in reading.

This is a transparent picture of a man whom his contemporaries and later ages regarded as a prodigious scholar but whose chief tools of scholarship seem to have been an abundant supply of books, index cards, paste, and scissors. In his extracting from earlier authorities without an awareness of their reliability Pliny appears to have been no different from other compilers, but in most other respects he was vastly superior to them. His energy and bodily self-discipline were boundless. He drove himself and his assistants to rummaging through books more efficiently than had any Roman before him, with the possible exception of Varro. He was ever sensitive to matters of human interest and had a flair for picking out the captivating tidbits from the barren tomes he was ransacking. It is not really science but the curious phenomena of natural science that absorb him. The keynote of his philosophy is that nature exists for man. Zoology and botany, the subjects of Books VIII–XXXII in the *Natural History*, interest him mainly for the boons and healing properties they provide for mankind and the hurts they may cause. Even his books on mineralogy (XXXIII–XXXVII) are focused upon man and afford Pliny an opportunity to present an interesting and valuable digression on the history of art. His diligent researches that encompassed practically all the visible wonders of nature, combined with a knack of selecting the fascinating bits of information, have produced a treasure chest of antique lore that has intrigued scholarly readers of every succeeding generation. J. Wight Duff, a leading British historian of Latin literature, calls Pliny's *Natural History* "one of the half-dozen most interesting books in the world."[1]

Two theories have been advanced to explain the remarkable unity and organization underlying Pliny's handling of his diversified subjects: (1) that he was following one or two Greek sources in large areas and that his other minor citations and borrowings were inserted here and there to impress his readers with the wide range of his research; (2) that his excerpts were so methodically arranged on index cards that when he was ready to write his topics fell into their proper places.[2] The mere accumulation of such a vast collection of information on a multiplicity of topics and the mustering of items in an orderly arrangement indicate that he had developed an efficient system of indexing his materials. It is probable however that in some areas like theoretical science, in which Pliny and his fellow Romans showed little interest, he was mainly depending upon one or two Greek sources, with perhaps Varro serving as an intermediary, and was introducing occasional statements of Roman authorities to bolster his compatriots' self-esteem and make it appear that they were not averse to exploring such fields. This practice also enabled him to fatten his list of authorities in Book I. He cites the emperor Augustus, for instance, in his list of authorities on cosmography, but his only reference to Augustus in that book is to his division of Italy into eleven districts—hardly a document. In general Pliny used Latin sources at every available opportunity—and these manuscripts were of course more readily procurable in Italy than those of Greek writers. Yet the number of Greek authorities he cites is more than double that of Latin ones.

No other writer in antiquity took the pains to list separately the authorities for each book in his work, Latin first (146) and then foreign (327), nearly all Greek. That his total list of 473 authors is considerably padded Pliny himself discloses in his Preface when he says that he has drawn from a hundred selected authors in compiling his encyclopedia. A page later he says he considers it a matter of pride that he has prefaced his work with a list of authorities. He regards it as a pleasant obligation and reflection of an honorable modesty to acknowledge the names of the writers to whom he is indebted. Many of his authorities, he has found, have copied word for word from their predecessors, without naming them, and this he feels is tantamount to theft. Pliny's hundred favorite authorities may be counted as his primary sources; and the 373 remaining sources listed in Book I may be presumed to be (1) authors whom he knew at second hand (he may have considered it scrupulous to credit the original author even when taking the citation from an intermediate source); or (2) authors from whom he had extracted isolated bits of information which he transferred to index cards for ready insertion at appropriate points.

There is no denying that Pliny got many of his citations of Aristotle at second or even third hand, but it is hard to follow the reasoning of a scholar who argues that all his Aristotle must have come through some Latin intermediary. Why should another Latin compiler have had the zeal to do reading that a man of Pliny's stamina and habits could not do? In isolated instances Pliny has been detected quoting verbatim from earlier Latin works that we possess, without acknowledging the author as his source. But can we be sure that he did not extract these quotations from intermediate sources and, in acknowledging his intermediaries, feel he had fulfilled his moral obligations? K. C. Bailey, in his excellent two-volume work on Pliny's chemical chapters,[3] wonders why Pliny, an honorable man in his eyes, appears to be continually paralleling Dioscorides' text but withholds any mention of him, listing instead as his own authorities the same nine sources listed by Dioscorides. It may be mistaken to assume that this is a case of Pliny's indulging in the common deception practiced by Latin compilers. It is known that Sextius Niger's lost work on *materia medica* was extensively used by Dioscorides, and Pliny frequently acknowledges his indebtedness to Sextius in these chapters.[4]

It is an appropriate fate that an encyclopedia as showily erudite and laborious as Pliny's has in turn engendered a spate of equally laborious and erudite dissertations that attempt to determine Pliny's relation to his sources. Experts in this field try to create impressive fabrics from bits of thread, a most optimistic kind of undertaking; and in classing Pliny with ordinary Latin compilers and treating him as such, they have given free rein to their enthusiasm for tracking down sources. Pliny divulges his relationship to his authorities, not as precisely as we would like according to our standards; but by ancient standards he was a man of unusual integrity and candor, and there is no reason for not taking him at his word.

Pliny's indefatigable curiosity about natural phenomena has led some to regard him as a fine example of an early scientific investigator; but when they go on to say that he had a scientific mind they fail to apply the rigorous criteria that modern usage of the word requires. The world of nature was to Pliny, like the world of books, hundreds of thousands of discrete phenomena, of which only thousands were interesting enough to be culled and catalogued in his notebooks. His avowed responsibility was merely to record these; only on rare occasions does he pass judgment on their credibility. Just as he did not usually distinguish reliable from inferior sources or trustworthy from worthless information, so he did not discriminate between ludicrous observations or deductions about natural phenomena and sound, penetrating ones. Just as he had a fondness for preserving from his reading items that aroused his curiosity, so he was interested in recording the more spectacular and amazing phenomena of nature. Like a child he stood in awe of nature and had little more than a child's sense of connection between cause and effect.

Time after time he makes some profound observation or draws an impressive inference (usually not his own), only to follow it immediately with some bit of lore that Frazer would compare to primitive folk beliefs. When he puts his curious anecdotes or fantastic tales in second place, after his more sober observations of nature and his efforts—albeit feeble—to interpret Greek theories, we get a feeling that Pliny may have had some slight power of discrimination; but there is nothing else in the text to indicate it. Today the scholarly world regards his encyclopedia as an important collection of documentary evidence—in the opinion of many, more valuable to folklorists as a repository of folk beliefs, which it makes available for comparison with those of contemporary primitives, than to historians of science.

It would be unfair to Pliny not to mention his compulsive urges to make independent investigations of natural phenomena. He draws attention on frequent occasions to his own researches or observations in the interests of science—often to confute some bookish Greek theorist. The climactic example is found in another letter from his nephew, written to the historian Tacitus. The senior Pliny was stationed, as admiral of the western fleet, at Misenum on the north end of the Bay of Naples, when Vesuvius began its catastrophic eruption over Pompeii and Herculaneum. Not satisfied with the view from Misenum, Pliny boarded a quadrireme and headed toward the volcano. With pumice stones and ashes falling about him, he stood on deck dictating to his secretary a detailed description of the rapidly changing and awesome sight. Landslides and a shoaling of waters forced him to turn down the coast to a friend's house, where he spent the night. He suffered a mysterious death the next day—perhaps from some cause directly connected with the eruption, more likely of an apoplectic attack resulting from the excitement. His death is certainly an example of martyrdom to the desire for scientific knowledge.

Such sporadic sorties into the realms and secrets of nature gave Pliny all the confidence he needed to stand up to the Greeks. Like Cicero and Seneca he patronizingly admits that the Greeks engaged in some remarkable theoretical speculations and calculations; but, on the other hand, such an extravagant undertaking as measuring the globe he thought sheer madness. Always the practical Roman can give the final answer to the theoretical Greek. Pliny exposes the ineptitudes of Greek scientists or comes to the support of their theories with his own examples. He lists some of the common proofs of sphericity of the ocean surface, such as drops of fluids taking on globular shape and ships disappearing from sight as they pass over the horizon; then he adds some of his own: that the surface of fluid in a goblet curves upward toward the center—hardly perceptibly, he admits—or that, whereas additional fluid poured into a full cup causes the surplus to overflow, twenty coins dropped into the full cup, because they lift the fluid into a peak, do not cause an overflow.

It is frightening to contemplate the results when such a headstrong, uncritical, and undisciplined mind as Pliny's addresses itself to Greek theoretical science—undoubtedly the field in which he is seen at his worst. After his opening book, containing bibliographical lists and indexes, Pliny devotes Book II to cosmography, the conventional beginning for compilers of treatises on geography. His list of seventeen Latin authorities for this book is obviously padded, Varro being a more important source than all the others combined. Among his twenty-seven foreign authorities we find Posidonius' name, a bit of evidence which supports the widely accepted theory that Pliny got his cosmography from Varro and that Varro got his from Posidonius.[5] Speculations about Varro's place in this tradition are materially assisted by a comparison of Pliny's cosmography with the cosmographical section of Martianus Capella's encyclopedia, compiled early in the fifth century. Striking correspondences between details and figures make it clear they were using the same ultimate source. Most scholars approve the view that Capella derived his cosmography from Varro. Although Capella's intellectual capacities and scholarship were far inferior to Pliny's, he shows himself much more patient than Pliny in trying to apprehend theoretical matters. He covers many more of the conventional topics of Greek handbook astronomy and gives a more comprehensive account—whence it is reasonable to conclude that Varro, too, was more at home in this field than Pliny, as he seems to have been in other theoretical subjects.

Pliny prefaces his cosmographical book with a rebuke for those who attempt to measure the earth or who are not satisfied with a single universe but speculate about an infinity of universes like ours. Oddly enough, at the close of the book he reports in admiring tones Eratosthenes' achievement in obtaining an earth measurement of 252,000 stades. He follows this with the solemn story of a certain Dionysodorus who left a signed letter in his tomb reporting that he had passed on downward to the center of the earth and had found the distance to be 42,000 stades; from this figure geometricians were able to calculate that the circumference of the earth was 252,000 stades. Pliny's geometricians were evidently using a gross approximation of three for π.

The body of Book II provides stock handbook information, interspersed with occasional personal reflections and frequent examples of folk beliefs. The universe is spherical and revolves in twenty-four hours. Pliny does not presume to know whether this rapid whirling of the celestial sphere and the swift movements of the planets produce harmonious sounds or not. The universe is constituted of four elements: fire at the top; next vital air, permeating the universe; suspended at the middle is the earth, together with the fourth element of water. The earth alone is motionless. The universe revolves from left to right; the planets have independent motions in the opposite direction. Ice-cold Saturn has the highest orbit and consequently that of longest duration: thirty years. Jupiter's orbit is considerably lower and is completed in twelve years. Mars, which has a ruddy glow because of its proximity to the sun, takes two years for its orbit. Pliny opines that Jupiter, situated between the extreme cold of Saturn and the excessive heat of Mars, enjoys a temperate condition. The sun, which exceeds the other planets in size and power, occupies a middle position among them and revolves in 365¼ days. It controls not only the seasons and things on earth but the stars and heavens as well. Subscribing to Stoic doctrines here as elsewhere, Pliny calls the sun the soul and mind of the universe and the ruling principle of nature. Below the sun comes Venus, shuttling back and forth with respect to the sun, called Lucifer when it is ahead of the sun and Vesper when it is behind. In a later chapter Pliny attributes this conspicuous shuttling to accelerations and decelerations resulting from anomalies in its orbit. Venus is so brilliant that it is the only planet (of course excluding the sun and moon) to cast a shadow. Its revolution is completed in 348 days and its maximum elongation is 46°. At its risings Venus emits a genital dew, stimulating conception of plants and animals on earth—a bit of the folklore Pliny collected. Mercury's orbit is still lower, requiring nine days less than Venus' for completion. Mercury's maximum elongation is 22°—later given as 23°. When discrepancies occur, as they often do, it is assumed that Pliny was following different authorities. Last in order comes the moon, a planet whose behavior is strange indeed. According to Pliny, Endymion was the first man to comprehend its puzzling phenomena. The moon's orbit takes 27⅓ days; after two days in conjunction with the sun, the moon is ready to begin another cycle.

Adhering to the traditional order of topics, Pliny next discusses the planets separately, beginning with the moon. The moon divides the year into twelve monthly parts. Its light is wholly derived from the sun; that explains why moonlight has a mild and gentle evaporating effect upon water, whereas the sun's rays draw it up forcefully. The moon is full only when it is directly opposite the sun; at other times it reveals to us only as much light as it receives from the sun. The moon, as well as the other planets, is nourished by water from the earth. The spots on the moon's surface at the first quarter show that it has not yet gained sufficient strength to draw up moisture with full force and is drawing up mud with the moisture. Pliny, like Plutarch, is attracted to popular beliefs about the moon and mingles folk fancies with Greek theory. With Pliny it is more fancy than theory. A valuable study could be made of the moon lore scattered through Pliny's encyclopedia and its survivals or parallels in folk beliefs today.

Pliny's discussion of eclipses in particular reveals how little patience or ability he had to comprehend Greek theory. He mingles remnants of demonstrations of Greek astronomers with incompatible Stoic doctrines. Pliny's only real interest in the subject is the wonder of it all and the obvious connection between eclipses and portentous events on earth. He has gathered from his reading that the earth and moon are the obstructing bodies in lunar and solar eclipses and that the earth's shadow is conical in shape. He proceeds to demonstrate that the shadow extends only as far as the moon, which it sweeps with its apex during lunar eclipses. The entire sun could not be eclipsed, he opines, if the earth were larger than the moon. (Greek astronomers had an approximately correct idea of the relative sizes of the earth and moon; according to Plutarch, the Stoics believed the moon to be the larger.) The huge size of the sun is indicated by the fact that shadows cast by a long row of trees extending several miles always fall parallel. Lunar eclipses prove the large size of the sun, just as solar eclipses prove the small size of the earth; this is a faint trace of Hipparchan observations and calculations that got into the handbook tradition. Lunar eclipses occur only at full moon, solar eclipses only at the conjunction of sun and moon. The cyclic period of eclipses is 223 lunations, but they have been known to recur within 15 days of each other.

The superior planets undergo their first stations at a distance of 120° from the sun (corresponding to Vitruvius' fifth sign, if, according to ancient practice, we count both ends) and their second stations at a distance of 120° on the other side. According to Pliny this means that Mars' stations last six months, while Jupiter and Saturn spend almost four months in each of their stations. Instead of offering a mathematical explanation of the planets' stations and retrogradations, as earlier detailed handbook accounts do, Pliny substitutes an astrological discussion. Since his strange aberrations about planetary exaltations and elevations are found to correspond, figures and all, with the statements made by Martianus Capella, it seems probable that Varro, their presumed mutual source on astronomy, derived this material from Posidonius, who was a believer in astrology and is known to have introduced it into his works.[6] According to Pliny, each planet reaches its point of highest elevation from the earth (apogee) in a certain sign of the zodiac: Saturn in Scorpio, Jupitur in Virgo, Mars in Leo, the sun in Gemini, Venus in Sagittarius, Mercury in Capricorn, and the moon in Aries. They also have points of greatest distance from their own centers (exaltation or absis) in still other signs: Saturn in the 20th degree of Libra, Jupiter in the 15th of Cancer, Mars in the 28th of Capricorn, the sun in the 29th of Aries, Venus in the 27th of Pisces, Mercury in the 15th of Virgo, and the moon in the 4th degree of Aries.

Pliny next takes up the planets' motions in latitude. All stay within the bounds of the zodiac except Venus, which transgresses by 2°; this explains why some creatures are born even on the desert wastes of the earth [!]. The moon ranges over the entire breadth of the zodiac, or 12°. Mercury's deviations amount to 8°, Mars' to 4°, Jupiter's to 3°, Saturn's and the sun's to 2°. Capella's figures for the deviations of these planets, in the same order, 12°, 12°, 5°, 5°, 3°, 1°, conform more closely to the Greek handbook tradition.

Pliny's explanation of the stations of the superior planets is a bizarre astrological one which does away with all the annoying complexities of epicycles and eccentrics. When the planet is struck by a triangular ray of the sun (i.e. at a point 120° from the sun), it is stopped in its course and driven upward by the powerful force of the ray. This elevation of the planet causes it to appear to be stationary. Then, in some strange manner, the powerful sunbeam gets in front of the planet and forces it to retrograde. He lets us believe that he has derived his theory of the motions of the superior planets from some earlier authority; but in the case of the inferior planets—a much more difficult set of problems, he confides—no one has attempted an explanation before Pliny himself. What follows, about the motions of Venus, is unintelligible jargon, calculated to impress his readers but not worthy of our attention here. After remarking shortly before (II.61) that Mercury's stations are too short to be detected, he now points out that its stations last almost four days. In saying that the sun has "four differences" (II.81), he may have been alluding to the inequality of the four quadrants of the sun's orbit; but he gives no figures for their duration, and goes on to mark the equinoctial and solstitial points at the eighth degree of Aries, Cancer, Libra, and Capricorn, points that coincide with those of Vitruvius' account.

On planetary distances Pliny is infantile. He baldly records that Pythagoras, "a brilliant man," ascertained that the distance from the earth to the moon was 126,000 stades (half of the commonly accepted figure, which he later records, for the measurement of the earth's circumference); that the distance from the moon to the sun was twice that, and from the sun to the celestial sphere three times as much. Pythagoras' figures, he adds, were acceptable to Sulpicius Gallus (a Roman astronomer of the Republican period). Pliny shows not the slightest interest in mathematics or in consistency. In his discussion of eclipses he had declared the moon to be larger than the earth and had spoken of the vast bulk of the sun compared to the moon and earth. If the reckonings ascribed to Pythagoras and Pliny's estimates of the sizes of the planets are combined, the result is a very crowded universe, practically a solid mass, consisting of earth, moon, sun, and sky, with no room for intervening planets—which, by the way, Pliny says elsewhere are all larger than the moon. This is the sort of thinking that was commonly put upon Pythagoras by admiring Roman compilers. Censorinus and Capella also record the estimate of 126,000 stades for the moon's distance in their discussion of the harmonious ratios in planetary intervals, and both presumably derived their material from Varro. If, as is commonly supposed, Varro was the main source of the astronomical portions of the works of both Pliny and Martianus Capella, it is possible that he too may have lacked regard for consistency. The corresponding discrepancies that are found in some of the figures of Pliny and Capella may also have been present in Varro's book on astronomy. Pliny held early authorities in reverence, as did other Latin handbook writers on science. He cites in support of his statement on distances in the heavens a Greek of the sixth century B.C. and a Roman of the second century B.C., both periods when science was in its infancy in those countries.

Although at the opening of this book he had disavowed interest in the subject of harmony of the spheres, Pliny now gives a brief account of the harmonious ratios of planetary distances—a regular feature in the Greek handbooks and probably so since the beginning, inasmuch as Plato had intrigued his followers by introducing the subject in his myth of Er at the end of the *Republic*. Greek handbook writers, as well as Roman, display an appalling ignorance of harmonic theory in handling these discussions in astronomical works; for example, in their musical scale the octave generally consists of eight intervals and nine notes. Some writers criticize their predecessors for assigning a note to the earth when it is stationary and therefore unable to produce a sound. The harmonious ratios are presumed to determine the distances of the planets from each other. Pliny assigns a tone to the interval between the earth and moon, a semitone to that between the moon and Mercury and between Mercury and Venus, a minor third between Venus and the sun, a tone between the sun and Mars, a semitone between Mars and Jupiter and between Jupiter and Saturn, and a minor third between Saturn and the celestial sphere. There are slight discrepancies in the intervals as given by Censorinus, Pliny, and Capella; since all three are believed to have derived their data from Varro, one scholar attributes the differences to discrepant manuscripts of Varro.[7]

Next follows a lengthy discussion of various phenomena and aspects of meteorology and physical geography, occupying one-third of this book. As noted earlier, Pliny's meteorological section bears marked resemblances, as well as a greater number of dissimilarities, to the discussion of the same topics contained in the works of Lucretius and Seneca; and it is generally believed that Posidonius was the leading authority of the three writers here, perhaps through transmission from Varro. The explanation of portentous phenomena in the heavens affords Pliny many fine opportunities to introduce folklore.

The last third of Book II is devoted to conventional topics of mathematical geography that are usually found at the beginning of geographical treatises. The earth is spherical and hangs suspended at the center of the universe, held in place by the force of breath which cannot escape from an enclosed world. Populations are distributed about the earth's surface. Antipodeans may well ask how we manage to stand on our heads. Pliny sees the acceptance or rejection of these views as a split between men of learning and the common herd. Dicearchus is cited here, as he is by other handbook authors, on the altitude of mountains, only to have his respectable estimate of his highest mountain (Pelion, approximately 6,000 feet) topped by Pliny's estimate of the height of some Alpine peaks as 252,000 feet. Accordingly, Pliny rejects Dicearchus' view that mountains are negligible features in comparison with the earth's mass, hardly noticeable protrusions on the earth's surface. That the surface of the oceans is spherical Pliny demonstrates by some good inherited proofs and, as we have seen, some faulty reasoning of his own.

In anticipation of his geographical survey, which attempts to cover the regions of the known world in the Greek manner of a coastal survey, Pliny offers the customary evidence that a continuous ocean surrounds all parts of the known world: the old stories of Macedonian sailors navigating the Eastern Ocean past the Caspian shores, of Indian sailors being swept by storms onto the northern coasts of Germany, and wrecks of Spanish ships being found in the Arabian Gulf. To these he adds recent confirmation: that Augustus' grandson was in charge of naval operations in that gulf when the figureheads of Spanish ships were identified there; and that a fleet dispatched by Augustus sailed past Jutland into an immense Scythian sea frozen by excessive moisture. When Pliny contemplates how human habitations are circumscribed by ocean, encroached upon by its great gulfs, and restricted on either side by the extremes of heat and cold, he is filled with pity for the mean lot of mortals, who have such overweening ambitions and so little scope for carrying them out. These echoes of sentiments expressed by Cicero belong to a stock of ideas that is common to both works and gives support to the theory that both men derived their views from Posidonius.

That the earth is at the center of the universe is proved by the equal hours of day and night at the equinoxes. Three celestial circles mark the variations of the seasons: the tropics of Cancer and Capricorn and the equator. The curvature of the earth's surface is proved by the fact that the appearance and elevations of constellations and conspicuous stars vary over different latitudes; the usual examples are cited—Canopus elevated in Egypt but barely visible on the southern horizon at Rhodes, and the North Star invisible in Egypt. For the same reason eclipses are observed at different times of the day in the east and west. In addition to the classic instance of the eclipse at Alexander's victory at Arbela being observed at an earlier hour of the day in Sicily, Pliny cites the acceptable recent example of an eclipse observed by the Roman general Corbulo in Armenia between the tenth and eleventh hours of the day and observed in Italy between the seventh and eighth hours. Ships traveling westward have more hours of daylight than those going in the opposite direction. Not to be overlooked is the report of one of Alexander's runners who, in covering 1,200 stades (over 100 miles) in a westward course, started out at dawn and reached his goal in the ninth hour of daylight, but in the reverse direction, although he was running downhill, did not return until the third hour of the night. (His speed on the outward course might have impaired his performance on the return run.)

The hours of daylight vary according to latitude: $12\frac{2}{3}$ equinoctial hours of daylight on the longest day of the year at Meroë in upper Egypt, 14 at Alexandria, 15 in Italy, and 17 in Britain. With the exception of the first, these are standard figures, introduced into the handbook tradition by Eratosthenes. But when Pliny comes to a fuller discussion of parallels, which he defers until the close of Book VI, he departs from Greek theory and offers instead what appears to be his own system of arranging lengthy lists of localities along each of seven circles.

He notes, in the discussion at the end of Book VI, that "circles" is the name preferred by Roman writers; the Greeks use the term "parallels"; but Pliny is badly confused. His seven "parallels," ranging in the east from southern India to the mouth of Caspian Sea and the tribes above it and converging in the west between Gibraltar and northern Spain, reveal how little he understood of Greek mathematical geography and how grossly he misconceived of the land masses of the known world. Three "parallels" run through Sicily alone. His first circle is marked by 14 equinoctial hours of daylight and his seventh by 15⅗, a difference of only one hour and 36 minutes. In between lie circles distinguishable by varying parts of an hour of daylight. On each circle he aligns many localities; whether appropriately or not is largely accidental. It is obvious that Pliny is once again parading a list of geographical names, many of them exotic. The system of Eratosthenes and Hipparchus of aligning localities along primary parallels or climata, varying by whole or half hours of daylight, is the sensible one—the procedure of a professional Greek cartographer. Above his seventh circle Pliny reverts to the Greek system of mathematical geographers, marking an eighth parallel from the mouth of the Don across to the North Sea provinces of Gaul (16 daylight hours); a ninth from the Hyperboreans to Britain (17 daylight hours); and a tenth parallel from the Rhipaean Mountains through Scythia to Thule. He refers to having pointed out earlier that Thule has alternate seasons of constant daylight and darkness. Actually, on two earlier mentions Pliny was unable to decide whether Thule had 24 hours of daylight at the summer solstice or six months of perpetual daylight in the summer.

After an inappropriately placed digression (II.191–211) on physical phenomena—earthquakes, emerging islands, and local wonders—Pliny turns to another usual preliminary feature of handbook geography, tidal theory. He accepts the Posidonian view that the moon influences the tides, which gives him an opportunity to introduce some lunar folklore about the waxing and waning of shellfish and of the blood content of humans according to the moon's phases, the putrefying effect of moonlight upon the corpses of animals, and its stupefying effect upon sleeping humans.

Still one more feature of mathematical geography precedes the regional geography: the dimensions of the known world. Pliny offers estimates of distances between places along two routes from India to Spain, with totals that differ by more than a thousand miles. Moreover, the total for each set of figures is incorrect; it is commonly assumed that his Roman numerals were badly handled by the copiers of his manuscripts. His next set of fifteen figures, for distances between places in a line stretching across the breadth of the known world from Ethiopia to Thule, is correctly added. Four of the figures are given precisely to the half mile. This occasional use of fractions is to create an impression of meticulousness, though at other times Pliny has no compunction about using round figures and rough estimates. His alleged authorities for his over-all dimensions were Artemidorus, who made a cruise around the Mediterranean and Black seas about 100 B.C. and based a periplus treatise in eleven books upon that voyage and upon earlier writers; and Isidorus of Charax, who was regarded as an expert on eastern geography and assisted Augustus' grandson Gaius on his campaign in Armenia. The complete text of Isidorus' *Parthian Stations*, an itinerary or roadbook from Antioch to India, still survives.[8]

Pliny devotes Books III–VI to regional geography. At the outset he has the usual misgivings of a compiler of geographical treatises about their readability and seeks his readers' indulgence: there is much ground for him to cover, many place names to be listed; he will be as brief as possible—which, we discover, means that he will do little more than list names of places and peoples,[9] inserting frequent scraps of information and occasional descriptive or reflective digressions of moderate length to relieve the monotony. For any reader, ancient or modern, to plod through the thousands of names would take the same sort of dogged but senseless determination that Pliny showed in compiling the lists.

The place names, mostly jumbled but sometimes arranged consecutively along an itinerary, give us a vague sense that we are progressing in a general direction; but the text, though it frequently states distances between places or islands, does not locate them with respect to each other—directions are occasionally given in locating islands, but rarely if ever along itineraries. A glimpse of Pliny's capacity for inflicting tedium upon himself and his readers may be gathered from his indexes in Book I, where he lists the authorities, topics covered, and total number of data, bits of research, and observations contained in each book. For Book II, on cosmography, he computes a total of 417 items, which he does not classify. For Books III–VI he offers summaries for seven classifications. Some kind soul, not impressed by the figures, has expunged them from the manuscript tradition for three of these books; but for Book VI they are preserved: towns 1,195; peoples 576; famous rivers 115; famous mountains 38; islands 108; extinct towns and peoples 95; grand total of facts, researches, and observations 2,214. We shall not quibble about an incorrect total in this ghastly summary in which Pliny takes obvious pride in the mere recording of famous rivers, famous mountains, and extinct towns and peoples. In his text he occasionally boasts that he has included names of towns not listed by earlier authorities. That he has a flair for selecting items of interest has been pointed out; there is evidence of it even in his geographical lists. But on the basis of his geographical books alone, Pliny deserves to be enshrined as patron of all writers of mechanical and obtuse dissertations.

In his regional scrapbooks are items of interest to the anthropologist: reports of fabulous creatures in Africa—men without heads; others with one, three, or four eyes; goat-pans, strap-footed men, dog milkers with dog heads—monstrosities that Pliny bequeathed to the Middle Ages and that persisted in illustrations of books on Africa down to the time of Livingstone and Stanley. There are also many details for the painstaking historian of geography who knows how to handle Pliny with caution and discernment. The description of Syria, for example, is the fullest extant account from antiquity and, when interpreted carefully, affords a great deal of information.

Pliny divides the known world into three continents, with the conventional boundaries of the Don and Nile rivers. Following the periplus practice, he begins at Gibraltar, ranges along the eastern coastal regions of Spain and Provence, then along the Mediterranean coast of Italy. At one point in his account of Spain he admits that he has selected memorable places or names merely because they are easy to pronounce in Latin. He makes frequent use of Marcus Agrippa's figures for over-all dimensions of provinces and countries, and praises him for the painstaking accuracy of his official survey. At other times Pliny shows no hesitation about criticising Agrippa and offering what he considers a more reliable figure. While he is moving about the western Mediterranean area, until he reaches the Istrian peninsula, Pliny makes a confident display of his scholarship, setting one authority against another, and drawing considerable information from official sources and recent campaigns. He frequently classifies towns according to official administrative terminology: colonies, municipalities, towns with Latin rights, free towns, federated towns, and tribute-paying towns. To the list of Alpine peoples he is able to add two new names, provided by the emperor Galba. Rivers break into the monotony of place names, and he often devotes more than a page to an important waterway. Periodically he stops to survey the islands embraced by the coastal regions he has just covered. After enumerating sixty-four islands in the northwestern Mediterranean, with special attention to Corsica, Sardinia, and Sicily, he resumes his course along the Adriatic coast of Italy. He takes pains to reproduce the complete text of a triumphal arch near Nicaea in Albania, and then follows it shortly with the conventional and convenient remark that the peoples round about are uninteresting and their names not easy to pronounce. There is a commendable observation, obviously not Pliny's own, discounting the persisting notion that a branch of the Danube empties into the Adriatic. Pliny suggests (as did Strabo) that the confusion originated in the report of Jason's return from the east by way of the Danube and the Adriatic, and points out that reliable authorities say the good ship *Argo* was portaged across the Alps. A trace of the erroneous notion that the Danube (Ister) River had a branch emptying into the Adriatic still survives in the name of the peninsula of Istria.

In the eastern Mediterranean Pliny has less self-assurance and follows his authorities (or authority) with very few expressions of his own opinions. From the shores of the Black Sea we are suddenly lifted over the Rhipaean Mountains and deposited on the shores of the northern ocean. Islands off the coast of Scythia are taken to be Baltic islands; and then we are on the shores of northern Germany. The account of Germany is shamefully scanty for a writer who served there as a cavalry officer and wrote a highly regarded history of the German wars in twenty books. Accounts of Britain and Thule follow, for which the antiquated Pytheas is his alleged authority, although the Romans had by this time subjugated nearly all of Britain. The Atlantic coasts of Gaul, Spain, and Portugal are surveyed as the northern periplus comes to an end. His scheme obliges Pliny to deal with Gaul and Spain in two parts.

We need not follow Pliny's southern periplus in detail across North Africa, east to Ceylon, and back around the southern shores of Africa. Latin authorities are again referred to frequently in the western areas. The Nile gets lengthy treatment; it was a source of fascination to all handbook geographers who included accounts of Egypt. The disproportionately detailed treatment of Syria has led some scholars to surmise that it was based upon his experiences as procurator of that province; but Pliny was a procurator in Germany and Gaul too, yet preferred to use the information of compilers, such as it was, on those countries. His attention to Syria probably stems ultimately from Posidonius' interest in Pompey's military prowess. Posidonius wrote a separate work on Pompey's eastern campaigns. Pliny does make occasional use of recent data obtained from military campaigns, pointing out that material furnished him by Corbulo corrects earlier writers on Armenia. Ceylon also receives extended treatment, a remarkable island to Pliny in many respects and a sort of Ultima Thule of the east.

As his southern periplus draws to a finish, the circumnavigation of Africa poses no problem to Pliny. Thousands of miles of coast line fade away in a mere difference of opinion: some authorities believe that the island of Cerne lies off Ethiopia, facing the Persian Gulf, while others maintain that it is at the southern end of Mauretania, across from the Atlas Mountains.

This is in outline Pliny's cosmography and geography—which furnish typical examples of his methods and materials. Here is the man who cast a spell over writers in the Middle Ages. If we have been inclined to deprecate or to be amused at the notions entertained by medieval writers, it will interest us to find parallels or sources of them in this book of wonders written by an esteemed colleague of a Roman emperor in the high period of the first century. The science of the Dark Ages had a spiritual kinship with Roman science from its very beginnings. The symptoms are clearly seen in Pliny: inability to comprehend Greek science or to distinguish between absurd anecdote and sober theory, between ungrounded opinion and brilliant original thinking. The written word is taken as sufficient authority. It is not by accident that the word "authority" comes from the same Latin root as "author."

VIII

SCIENCE IN THE SECOND CENTURY

The second century presents a virtual hiatus in the course of Roman science and affords us a fine opportunity to pause and gain some perspectives. It is an arresting thought that the heyday of Roman science—such as there was of it—had now passed. This malformed babe, born into an inhospitable world, allowed to develop rickets from lack of proper nourishment, had fallen into a state of decline without ever having given indications of attaining manhood. To us Varro, Vitruvius, Celsus, Seneca, and Pliny are adept compilers; but these were the great authorities who were to fill medieval and Renaissance men of science with awe and veneration; these were the figures which, in gravely contemplative attitudes, were to decorate rotunda columns in the frontispieces of early folio editions of scientific literature. There would be matching pedestals for the busts of Greeks and Romans—for the men of science as for the poets, orators, and philosophers. Greece had its Homer, Demosthenes, and Aristotle; Rome its Virgil, Cicero—and Pliny. Renaissance intellectuals seemingly could not appreciate the difference between scientific genius and the réchauffés of an encyclopedist. To Petrarch, for example, Varro was "the third great Roman luminary,"[1] ranking only after Cicero and Virgil. Varro and Pliny were unquestionably Rome's greatest authorities on scientific matters but they were actually erudites who had less interest in theoretical science than in any other subject.

With classical Roman antecedents such as those we have discussed, what prospect of development did science have in the Latin West for the next millennium, deprived of an understanding of the true character of Hellenistic science and cut off from later scientific research at Alexandria and other centers in the East? The writings of Varro and Pliny would be transmitted, their doctrines and data becoming less intelligible and accurate each time some Latin writer incorporated them in a new compilation. Varro's theoretical science was to experience a revival of a sort in Capella's encyclopedia, through which traces of it would survive; but Varro's book itself was soon after to pass out of sight. The manuscripts of Pliny's bulky *Natural History* were to be copied and treasured in libraries throughout the Middle Ages—a testimonial to Pliny's knack of accumulating amazing and incredible tidbits and to their attractiveness for medieval readers. There would be a revival of Pythagoreanism in the second century, and new Greek rehashings of these antique but still intriguing doctrines of the master. One Neopythagorean compilation, by Nicomachus, as we shall see, would be translated twice into Latin and provide the Middle Ages with materials for manuals on arithmetic and harmony. There would be a revival of interest in Platonism as the new philosophical religion of Neoplatonism swept to Rome and won many converts among literati. Neoplatonism was to focus attention once again upon Plato's *Timaeus* and, through the commentaries of Chalcidius and Macrobius, would give medieval readers the impression they were encountering the best productions of Greek science. Lastly there was the amazing figure of Boethius, who in the sixth century was to conceive a grand scheme of commenting upon and translating into Latin all the works of Plato and Aristotle, together with the mathematical works of Ptolemy. If this gigantic undertaking had been feasible for one man and had been carried out it might have made possible in the West a true revival of Greek science. The set of translations and commentaries contemplated by Boethius, if completed, would have radically changed the intellectual history of the Western world in the early Middle Ages; but unfortunately Boethius' life was cut off early, before he got deeply into the scientific works of his project.

When the masterpieces of Greek scientific literature finally became available in Latin translations in the late Middle Ages and Renaissance, the reverence for the ancient Roman compilers continued to be manifested. Even a brilliant scientist like Copernicus expresses admiration for the ingenuity of a Roman encyclopedist and discusses his views next after those of Ptolemy. The mere fact that an author belonged to an early age made him a weighty authority. Latin masters of science were as worthy of respect as the Greek masters; equal representation was given to both in the gallery of learning. And so lavish folio volumes were printed containing the writings of Celsus, Vitruvius, and Pliny, ready to be placed on library shelves alongside editions of Galen, Hippocrates, and Ptolemy. Frequently printed editions of Latin scientific authors antedated those of Greek authors in the same field. Even a wretched work like Solinus' fourth-century *Collectanea*, to be discussed at length in the next chapter, had two editions and five impressions before 1500, an Italian translation by 1557, a Spanish translation by 1573, and an English translation by 1587. The sort of deference Renaissance humanists showed toward the Roman compilers may be found persisting even to the present. Some recent historians of science have found many complimentary things to say about the Romans.

Science is a sequestered occupation of gifted minds. Under favorable conditions of research, when keen individuals are stimulated by each other's discoveries or reasoning, or are sparked by the ideas of predecessors, brilliant results are achieved. Because it flourishes in seclusion, scientific research may not reach its apogee at a time when the best literature and works of art are produced; ancient Greece and our contemporary age provide eloquent testimony to this. In Greece science reached the peak of its development in the Hellenistic and Roman imperial periods, long after the golden age of Greek literature and art.

With the lay handbook science of antiquity the case was entirely different. Such compilations were mere book learning and inevitably reflected the rises and declines of intellectual life and of letters in general. Thus it was possible, in the Mediterranean world of the second century, for theoretical science and medical research to experience a high-level revival at Alexandria, at the very time when Latin handbook science was markedly deteriorating in the West.

The sharp decline in Latin letters in the second century was accompanied by a correspondingly sharp drop in the level of scientific writings and interest. Students of Latin literature, recalling that after the close of the Silver Age (A.D. 18 to A.D. 138) the outstanding men writing in Latin were such unimposing figures as Aulus Gellius, Fronto, and Apuleius, would not expect to find significant merits in the Latin scientific writing of the time. The waning interest in Latin literature is seen in the growing fondness for epitomes and in the fact that many intellectuals were abandoning the vernacular and writing in Greek. The emperor Hadrian set the fashion, early in the century, and Suetonius and Fronto followed suit, writing in Latin or Greek according to their whim. The emperor Marcus Aurelius composed his *Meditations* in Greek.

Attention should be drawn however to a few manifestations of scientific interest among Latin writers of this century. Suetonius (c. 69–140), last of the Silver Age writers, known to us for his racy *Lives of the Twelve Caesars*, was in his own day regarded as an antiquarian and man of great learning. Little is known about his lost *Pratum* ("Meadow"), a sizable encyclopedia which included separate books on the universe, zoology, and botany, among other subjects. Isidore of Seville, five centuries later, made extensive use of this work. Aulus Gellius (c. 123–165) made a large collection of miscellaneous scraps of information entitled *Attic Nights*, which contains little of scientific interest but throws much light upon the habits of compilers of this and later periods. This work, surviving almost intact, shows no semblance of order and no critical or literary ability on the part of the author; it is simply a bundle of gleanings from earlier writers. Contemporary works are not credited, but the authors are introduced as speakers of their quoted words. Gellius cites numerous early writers, but it is commonly believed that he went directly back to no one earlier than Varro. At one point he pretends to have derived some engaging tales from six little-known authors; later he admits finding them in the seventh book of Pliny's *Natural History*. Our only interest at this time in Gellius' miscellany is that he has preserved some extended quotations on Neopythagorean arithmetic from Varro's encyclopedia of the liberal arts.

Still another writer to interest himself in popular science was Apuleius (born c. 125), who is best known today for his highly readable novel, *The Golden Ass*. Apuleius produced a free translation of a treatise *On the Cosmos*, falsely ascribed to Aristotle but probably written in the first century B.C. or A.D., and a translation of Nicomachus' *Introduction to Arithmetic*. Both translations had an appreciable influence upon scholastic writers during the Middle Ages. *On the Cosmos* is a compendium only thirty-three pages in length. The scientific portion (chaps. 2–5) recapitulates conventional handbook doctrines about the spherical nature of the heavens and earth, the motions of stars and planets, the location of land masses, seas, and gulfs, and some details of various meteorological phenomena. W. Capelle argued the Posidonian character of the treatise; J. P. Maguire differed with that view, stressing the reliance of the author upon Neopythagorean writers.[2] By the time this treatise was compiled Neopythagorean, Platonist, Aristotelian, and Stoic tradition had become confused, and popular writers were representing eclectic and not sectarian positions.

At Alexandria, in the second century, Greek scientific studies were reaching another pinnacle as the two towering figures of Greek science of the Christian Era, Ptolemy and Galen, brought their respective fields of astronomy and geography and of medicine to their highwater marks in the ancient world. Both men, to be sure, were able to reach such marks because they stood upon the shoulders of predecessors who lived in the heyday of Hellenistic science, Ptolemy on the shoulders of Hipparchus, and Galen on the shoulders of Herophilus and Erasistratus. But both men acknowledged their heavy indebtedness and enable us, from their free use of quotations and from careful accounts of the procedures and calculations of their Hellenistic precursors, to make partial reconstructions of the research done at Alexandria and Rhodes during the second and third centuries B.C. Ptolemy and Galen expertly incorporated the brilliant work of their predecessors into their own systematic bodies of knowledge and contributed highly significant additions and refinements to their inherited findings. Each was a master of style and arrangement of materials. As a result the works they produced became the awe and despair of all emulators in these fields. To these apparently perfect systems of scientific knowledge hardly anything significant, it seemed, remained to be done, either by way of addition or revision—though to be sure there was still scope for the commentator's zeal and acumen. Ptolemy and Galen bathed in the warm, late-afternoon sunlight of antiquity, shortly before the sun was to set upon the ancient world. Their popular reputations were based on the fact that, in addition to being outstanding researchers in their specialties, they were also extraordinarily gifted in many fields. In this century men of impressive intellectual attainments received lavish fees for spellbinding audiences from public lecture platforms—as the intellectual world crumbled about them.

We should note here that although the body of the great work on astronomy which Ptolemy entitled *Syntaxis mathematikē* ("Mathematical System"), but which is more generally known as the *Almagest*, was essentially a mathematical demonstration to account for the positions and motions of celestial bodies, his assumptions in Book I about the earth and its relation to the physical universe resemble most closely the views expressed by Aristotle in his treatise *On the Heavens*. Ptolemy rejects the Heraclidean theory of the diurnal rotation of the earth. He uses empirical arguments to substantiate conventional doctrines of a limited and spherical universe which makes diurnal rotation around a spherical, motionless earth situated at its exact center, and of the earth being a mere point in comparison with the boundaries of the universe. Some experts, like Duhem and Heath, have taken the extreme view that Ptolemy was merely presenting a geometrical account of celestial motions and did not intend to represent physical phenomena with precision. It is true that he was more keenly concerned about mathematical accuracy than about having his calculations coincide with observations; but it cannot be said that Ptolemy ignored what he believed to be physical realities.

A similar attempt to achieve a reconciliation of physical and mathematical data and to handle his subject on two different levels is found in Ptolemy's other *opus magnum*, the *Geography*. The exact converse of the case of the *Almagest* is presented here. The introductory portion (Book I) is a superb exposition of the mathematical principles of cartography—the methods of determining the latitude and longitude of places by astronomical means and of representing a spherical surface on a plane. Then, instead of applying his mathematical principles, Ptolemy proceeded to base the body of his work (Books II–VII) upon shoddy data accumulated mainly from the dead reckoning of travelers on land and sea. Later ages were mistaken in supposing that the *Almagest* was a physical system and the *Geography* an application of mathematical principles to cartography. These misapprehensions may have arisen, as they often did in antiquity, from erroneous impressions created in the opening chapters of each work.

Other parallels between the physician Galen and the astronomer and geographer Ptolemy could be pointed out; but one in particular is important to us as we seek to trace the connections between Greek handbooks and Latin science. Why did some knowledge of the researches of

Eratosthenes, Hipparchus, Herophilus, and Erasistratus filter into the Roman encyclopedias of Varro, Celsus, Seneca, Pliny, and others, while the works of Ptolemy and Galen remained virtually unknown to the West for a millennium?[3] Certainly their complexities and abstruseness do not provide the answer; the studies of those four predecessors were equally abstruse, and yet smatterings of information about them circulated in Italy in popular writings which preserve for us bits of knowledge about them we would not otherwise have had.

The answer to the question is found in the Greek handbook. Some notion of the researches of the early Hellenistic scientists had been provided for Greek readers by compilers of popular compendiums and then transmitted to Latin readers, more often than not by way of Posidonius and Varro. But if popular accounts of the work of Ptolemy and Galen were prepared and circulated in the eastern Mediterranean world, they were not transmitted to the West at that time or in the centuries immediately following. Ptolemy's *Syntaxis mathematikē* became the subject of commentaries by eminent mathematicians—Pappus, Theon of Alexandria, Hypatia, and Proclus—and that tradition went eastward, to be developed by the Arabs. Galen's works were handled by Byzantine physicians—Oribasius, Aetius, Alexander of Tralles, and Paul of Aegina— and then went farther eastward, to the Syrians and Arabs.

If some slight knowledge of the achievements of the early Hellenistic scientists had not been circulated through popular handbooks, Eratosthenes and Hipparchus would have been mere names to classical Latin writers, as they were to become to writers in the Dark Ages. In the West Ptolemy and Galen remained mere names, known almost solely by reputation until translations of their works from Arabic and Greek texts became available between the twelfth and fifteenth centuries. When historians of science categorically refer to Ptolemy's *Almagest* and *Geography* as the bibles of astronomy and geography and to Galen's definitive corpus of medical writings as the bible of medicine for fourteen centuries,[4] we must not be misled into supposing that these authors had any appreciable influence upon Latin scientific writing during the first millennium of the Christian Era.

Further confirmation that the Greek handbook was the deciding factor in determining whether or not Greek scientific research would become known to Latin readers is furnished in Neopythagoreanism—the next development to be considered here. But before turning our attention to the tenets of this school, it will be helpful to make a few brief remarks about the commonly shared doctrines that have been and will be designated as Pythagorean, Neopythagorean, Platonic (or Platonist), and Neoplatonic. The elements of Pythagoreanism that concern us in this volume are harmonic theory, planetary theory, and arithmetic, the very elements that Plato deals with in the *Timaeus*. When Platonist writers of the Hellenistic or early Roman imperial periods handle the doctrines, we refer to them as Platonic. The Pythagorean school passed out of existence as an organization in southern Italy late in the fourth century B.C., but some of the doctrines continued to be maintained or adapted by Platonists, Aristotelians, and Stoics, as well as by Pythagoreans. In the first century B.C. Pythagoreanism experienced a revival, at Alexandria and Rome, from which time it came to be known as Neopythagoreanism. Similarly Platonism, with substantial admixtures of Neopythagoreanism and the tenets of other philosophical schools, developed into Neoplatonism—a term used to designate the revival of Platonism after the middle of the third century of the Christian Era. Writers belonging to any of these schools are found freely embracing or copying Pythagorean-Platonic doctrines. How these doctrines are labeled is determined by the period in which they were written and the sectarian affiliations of the author.

By A.D. 100 the doctrines of Pythagorean arithmetic and arithmology (attribution of secret powers to the numbers of the sacred Pythagorean decad) should have been very stale. They had been circulated in one form or another, with little change, for over five hundred years. To the ancient Greeks arithmetic was a philosophical study, not to be confused with mere reckoning in the market place and at home. From the time of Socrates, and perhaps even of Pythagoras, educated Greeks had poked fun at people who thought of it as a practical art of computation. Serious students of mathematics observed two distinct divisions of arithmetic, originating among the earliest Pythagoreans: (1) the theory of numbers, a sober discussion of the relationships of numbers; and (2) the "theology of numbers," or arithmology, an appreciative recital of the mystical and even magical properties of numbers. As one might expect, number lore *why ?* was more fascinating to the ancients than plain theory. Each number of the decad was revered by Pythagoreans from earliest times and was given epithets testifying to its divine and magical powers and associations. Treatises devoted wholly to discoursing upon the attributes of the numbers from one to ten appeared in profusion, most of them compiled by men reputed to be members of the Pythagorean sect, some by men of high intellectual stature.

In such treatises each number is elaborated upon until the writer's imagination (and the modern reader's patience) is exhausted. A few compressed examples will serve to indicate their nature. One is not itself a number but is the source of all numbers; it is male and female, odd and even, and associated with the Godhead. Four is the first solid number; it is the first number to have two means and accordingly was used by the Creator to bind the four elements of the universe. Six is the first "perfect" number, being equal to the sum of its aliquot parts, one, two, and three. It is also the product of the first male number, three, and the first female number, two, and is therefore sacred to Venus. It refers to the six motions—upward, downward, to right and left, and forward and backward—and to many other combinations and relationships. (Some writers preferred to associate the motions with the number seven, adding rotary motion. Plato was responsible for this confusion, referring in *Timaeus* 43b to the six motions and in 34a to the seven motions.) Seven is found to be intimately associated with man, from the stages of development of the embryo in the womb, through marked changes that occur every seven years during early life, to the critical climacterics that occur in mature years; seven controls both the moon's periods and the menstrual period of women; seven marks the number of man's vital organs, of his major vascular ducts, of his tissues, of the visible parts of his body (head, trunk, two arms, two legs, and the *membrum virile*), and of the openings in the body (including two nostrils but not the excretory apertures). The numbers eight, nine, and ten, like seven, get extended treatment, in keeping with their remarkable attributes. In such obvious forcing of objects and critical events into schemes of numbers—the beard appearing in the third heptad of years, for instance—it is not surprising to find discrepancies, like the assignment of the first growth of beard to the second heptad by some writers.

Scholars comparing a dozen or more of these treatises at a time, by placing the corresponding passages in parallel columns, have found extended passages of a page or more—in one case six pages—of verbatim copying. It is clear that these were sectarian, quasi-religious materials that were transmitted with little change from an early period of the Pythagorean school down into the Middle Ages. The Neoplatonist school of philosophy absorbed much of this number lore and associated the number one with the Supreme and Ineffable God, commonly referred to as "The One."

The quickening of interest in Pythagorean number theory and number lore during the second and first centuries B.C. was largely the result of the fascination that Plato's *Timaeus* and *Republic* held for Greek readers at that time. Handbook and commentary authors, in attempting to explain the cryptic mathematical passages about the World-Soul in the *Timaeus* and the nuptial number in the *Republic*, naturally resorted to Pythagorean arithmetic, since Plato was presenting Pythagorean doctrines in those works. Posidonius, during the first half of the first century B.C., embraced Pythagorean number theory and lore in his works and made the subject still more fashionable. After him *Timaeus* commentaries gave permanent place to Pythagorean arithmetic. Varro next introduced Posidonian materials into the books on arithmetic, geometry, and music in his encyclopedia. Nigidius Figulus, a contemporary of Varro and considered by Aulus Gellius to be next to Varro in learning, also had much to do with the growing interest in Pythagoreanism at Rome, according to a statement of Cicero. Thereafter, during the period of the early empire, Latin compilers dabbled generally in Pythagorean arithmetic.

Then, if interest in the subject was beginning to wane, it was revived by two works which were to have a far-reaching influence upon science through the Middle Ages. The first of these was the *Introduction to Arithmetic* by Nicomachus, perhaps the best separate treatment that had ever been given to the theory of numbers by a Greek compiler, and a fine example of Greek skill in preparing a synthesis of a discipline. The second was the *Manual of Mathematical Knowledge Useful for an Understanding of Plato*, compiled by Theon of Smyrna, which has already been discussed with other Posidonian handbooks. Substantial portions of this were translated and paraphrased in Chalcidius' *Commentary on the Timaeus* (c. A.D. 325), and through that work Theon's doctrines were widely circulated in the Latin West, particularly in the twelfth and thirteenth centuries. The close resemblance of the two treatises that F. E. Robbins has traced [5] makes it useful to keep Theon's mathematical section in mind while analyzing Nicomachus' *Introduction*.

Nicomachus, who was born at Gerasa in Judaea and flourished about

A.D. 100, wrote handbooks on Pythagorean arithmetic, geometry, number lore, and musical theory, all designed for lay readers. Although almost nothing is known about the development of arithmetic in the four preceding centuries, it is safe to assume that neither Nicomachus nor Theon was a creative mathematician; but Nicomachus, like other ancient compilers, made occasional pretenses to expert knowledge and claimed new discoveries. The bulk of his arithmetic appears to have been as old as the early Pythagoreans. His success as a popular writer was so great that his reputation in subsequent ages until the Renaissance rivaled and sometimes surpassed that of Euclid.

Nicomachus wrote a treatise on arithmology, entitled the *Theology of Numbers*. Though it has not survived, its contents are known to us from another treatise, erroneously ascribed to Iamblichus, substantially reproducing its matter. And the Byzantine scholar Photius wrote an analysis of Nicomachus' work. With this, enough has been said about these curious documents of which Nicomachus' *Theology* was one. Let us pass on to his more important work on arithmetic, one in which he also reveals a reverential attitude toward Pythagoreanism.

Book I of his *Introduction to Arithmetic* opens in the conventional manner, devoting six chapters to traditional protreptic arguments on behalf of the pursuit of philosophical and pure mathematical studies. One suspects that Nicomachus has obtained his arguments from other handbook writers and not from original sources, for his quotations of Plato are never accurate and sometimes have the same discrepancies as Theon's. Theon too opens his handbook with arguments for the study of mathematics and quotes the same two passages from Plato's *Epinomis* and *Republic*, in the same order as Nicomachus. It has been suggested that Nicomachus and Theon were independently quoting the gist of Plato's words, loosely from memory.[6] More probably both were using handbook sources, and their discrepancies in quotations and presentation of arguments of the early masters can be attributed to careless transmission of traditional materials by successive generations of compilers. We cannot rule out the possibility that both men read Plato's original works and had them readily accessible, but it is unlikely.

In chapters 7–12 Nicomachus deals with the classification and properties of numbers and their division into odds and evens. Even numbers are subdivided into even times even, odd times even, or even times odd. Theon offers the same classification. Odd numbers are subdivided into prime and uncompounded, compounded, or compounded but prime to each other. Chapter 13 explains the so-called "sieve" method of Eratosthenes, used to find prime numbers. It also presents the rule for determining whether any two given numbers are prime to each other or not, handling the Euclidean problem of the greatest common divisor in words, without using straight lines or symbols. Chapters 14–16 define excessive, deficient, and perfect numbers. Chapters 17–23 discuss the relations which numbers bear to other numbers. Nicomachus offers an elaborate classification of numerical ratios which are greater than unity and their corresponding ratios which are less than unity. Chapter 23 sets forth the rules for deriving the various forms of inequality of ratio from each other. Theon attributes this method to Adrastus. It seems reasonable to suppose that Eratosthenes also knew the method and that he and Adrastus both presented it in their lost commentaries on the *Timaeus*.[7]

Book II opens by discussing a converse handling of the rules developed in chapter 23 of Book I. Theon also offers the converse, attributing it to Adrastus. Chapters 3–4 present a method for deriving the series of superparticulars, discussed in chapter 19 of Book I, from the successive series of multiples. Chapter 5 deals with other ratios produced by combination of ratios. Chapters 6–20 are devoted to plane and solid numbers. Nicomachus begins by pointing out that numbers may be classified as linear, plane, solid, cubical, and so on; that numbers really underlie geometrical figures as they underlie everything in the universe, and that arithmetic is therefore more basic and elementary than geometry. The number one represents a geometric point. The number two, or any number exceeding it, may represent dots or units juxtaposed in linear arrangement to form a line—the first dimension. Triangular numbers begin with three, which may be represented by three dots or units juxtaposed but not in line, forming a triangle—the second dimension. Similarly four and numbers exceeding it may represent the tetrahedron and other solid figures.

These doctrines about figured numbers appealed to Platonist and Neoplatonist commentators and compilers, even if they were unable to comprehend the rest of Pythagorean number theory, as the proof and revelation that numbers were the key to the understanding of the universe. The number one referred to the Neoplatonic Supreme Deity. It must have been because four is the first solid number, they reasoned, that it was used by the Creator in fabricating the World-Soul in the *Timaeus.*

In successive chapters Nicomachus deals with triangular, square, and pentagonal numbers, arranging them diagrammatically in the text, by way of illustration; then he passes to hexagonal, heptagonal, and other polygonal numbers, listing in separate columns those that belong to each class. With chapter 13 he takes up solid numbers, classified as pyramids, cubes, beams, tiles, wedges, spheres, and parallelepipeds. The remainder of the book, from chapter 21 on, deals with the definitions and classifications of the various kinds of proportions.

The virtue of Nicomachus' *Introduction,* according to historians of mathematics, is that it provides the most comprehensive extant account of the elements of Greek arithmetic. Theon demonstrates that he was a better mathematician than Nicomachus, but his materials are poorly arranged and explained too concisely. Correspondence between the treatments of Theon and Nicomachus and the arithmetical material contained in Books VII–IX of Euclid's *Elements* is close enough to indicate that this portion of Euclid was treated like a separate handbook of arithmetic and absorbed into the handbook tradition. There is a great deal of difference, though, between the methods of Euclid and Nicomachus. Euclid always offers proof of his propositions; Nicomachus never does. That proofs had been dispensed with in the handbooks is quite in keeping with their rudimentary and simple character. Euclid's numbers are represented by lines with letters attached, and the lines do not necessarily have specific values; Nicomachus does not designate numbers by lines, so he has to resort to circumlocution to distinguish between undetermined numbers.

The *Introduction to Arithmetic* was translated into Latin about A.D. 160 by Apuleius. Cassiodorus and Isidore of Seville refer to this translation, but it has been completely lost. Boethius' *On the Elements of Arithmetic,* published in the sixth century, is a free translation of the *Introduction.* In this form (and hence known to the Middle Ages as "Boethian arithmetic") Nicomachus' handbook immediately became a standard work in the Latin West, along with Euclid's *Elements,* and remained so for more than a thousand years. The Arabic world also became acquainted with Greek arithmetic through a translation of the *Introduction* by Thabit ibn Qurra in the ninth century; parts of this exist in manuscript.

After the publication and translation of the *Introduction*, arithmetic began to rival or supersede geometry as the leading mathematical subject in school curricula. This was particularly true in the Latin schools, where the practical form of arithmetic, the art of computing, had had no difficulty in justifying its place in the elementary curriculum. Moreover, the usefulness of a knowledge of arithmetic to Christians in calculating the dates of Easter and other movable feasts was to add importance to its study. After Nicomachus' death Euclid continued to hold his position as the supreme authority in geometry, but henceforth he had to share his domination of mathematical studies with Nicomachus, who was regarded as his compeer and the master of arithmeticians.

An indication of the esteem in which Nicomachus' work was held in later centuries is found in the number of commentaries that have survived or are known to have existed. Iamblichus' commentary, the most important of the lot, is a comprehensive treatise on arithmetic of about A.D. 300, based upon Nicomachus' work. Those of Proclus Procleius and Heronas, from the fourth or fifth centuries, have been lost. Commentaries by Asclepius Trallianus and Johannes Philoponus survive as collections of scholia in the manuscripts. The commentaries of Iamblichus and Philoponus have been published in modern editions, as have other scholia by Soterichus and anonymous scholars. Lastly, a commentary was prepared by Camerarius in the sixteenth century.

It should be pointed out in concluding this chapter that Plutarch, the greatest Greek literary figure of the second century, is an important source of knowledge about the Greek handbook movement. His *Moralia* is a vast miscellany, comparable in size and scope to Pliny's *Natural History*. Many of the doctrines it records are commonplaces collected from popular handbooks. Plutarch demonstrates greater aptitude for assimilating and reporting scientific information than Pliny, but he is no less credulous and uncritical of quaint and incongruous data. In his essay *On the Face in the Orb of the Moon*, for example, he alludes to Aristarchus' heliocentric theory, gives Aristarchus' figure of the distance from the sun to the earth, reports a minimum estimate of the moon's distance as fifty-six earth-radii, a respectable figure—and in another passage ridicules the idea of a spherical earth and antipodes. If the present volume were a study of the Greek handbook movement, the *Moralia* would receive considerable attention; but Plutarch did not influence the Latin handbook movement.

IX

THIRD- AND FOURTH-CENTURY COSMOGRAPHY

Cosmography was the chief interest of Latin scientific writers during the closing centuries of the western empire. As we examine their works our thoughts inevitably return to Varro and Pliny, the great Roman authorities in this field. Beginning with the third century we shall observe the influences that these two men exerted and be led to reflect upon the vicissitudes of their writings in the West—how Pliny's encyclopedia survived and Varro's encyclopedia was allowed to disappear.

We would regard Varro as clearly superior to Pliny as a man of learning. Varro's reputation among his contemporaries and his output of books could not be matched by Pliny. Yet Pliny, living at the height of the Roman Empire, would certainly not have agreed with our judgment. In proudly displaying, at the opening of his work, his list of nearly five hundred authors, including Egyptian, Babylonian, Persian, Etruscan, Carthaginian, Greek, and Roman authorities, he undoubtedly thought of himself as standing on a pinnacle, looking down upon the entire range of human knowledge. To his compatriots knowledge was merely the learning and human experiences accumulated through the ages and recorded in books; and Pliny, by reason of his abundant store of physical energy and his rigorously disciplined habits of study, was in a better position than any man before him to encompass and extract the learning and experience of all civilized nations—the Roman naturally being the flowering of the others. So it seemed to Pliny and his colleagues that he could rightfully lay claim to being the most learned man of all time.

In matters of theoretical science Varro's superiority to Pliny is even clearer. Pliny expressed his lack of interest in such subjects and whenever he handled them revealed, by his glossy and ambiguous style and his anecdotes and amusing illustrations, his inability to comprehend the Greeks. Varro, on the other hand—on the basis of the assumption that the astronomical portion of Capella's encyclopedia was largely derived from him, as well as on other grounds—appears to have been able to absorb the basic elements of Greek theory.

We do not know exactly where Pliny stood in the rise and decline of Roman interest in science. His relation to the scientific compilers of the classical age is not clear because nearly all of their works have been lost. But in his relation to posterity Pliny may be considered a leading precursor of the Dark Ages. A number of representative works of Latin science are extant from the third century and later, in which we are able to observe the gradual stages of the decline.

It may seem paradoxical, on first thinking about the matter, that the scientific writings of Varro have perished almost completely, whereas the inferior compilations of the later Roman Empire, including those derived from Pliny, have been preserved. The explanation lies in the fact that science during the Dark Ages was a continuation, in a deteriorated state, of the scholarly attitudes and practices of classical Roman science. Varro's encyclopedia was too theoretical for the vast majority of Roman intellectuals; when it was made more palatable in digests and adaptations, the original was unable to withstand the competition and dropped out of circulation. Pliny's *Natural History*, being a book of wonders and more engrossing than the austere discipline of Varro's quadrivium, was able to survive intact even though it too was fragmented, digested, excerpted, and adapted in many forms. As time passed, readers had less and less desire to study elaborate or systematic presentations of classical science and preferred to use concise compendia. The later compilers and digesters managed to give an impression of mastering their subject while skimming the surface; consequently they had a greater attraction for medieval scholars and their works were circulated in greater numbers. Moreover, the persistent tendency of handbook readers to prefer compilations of a more recent vintage made the latest of the classical compendia the most popular and enduring.

Such a compendium was the book of wonders compiled by Solinus, largely by distillation from Pliny's encyclopedia. About Gaius Julius Solinus almost nothing is known. He seems to have lived in the third or fourth century. It is clear that the book was written before 400; it has been suggested that his style belongs to the third century. On the other hand, the arguments that have been advanced for placing Solinus' dates before the emperor Constantine must be regarded as inconclusive in the light of what is now known about the habits of handbook authors. That he refers to Byzantium instead of Constantinople has no bearing here since he was paraphrasing Pliny; nor does the fact that he makes no mention of the divisions of provinces created by Diocletian and Constantine.[1] We would not expect a compiler to keep up to date. The fact that there is no reference to Christianity in Solinus' book is not the slightest justification for assigning an early date to it. A long list could be drawn up of fourth- and fifth-century pagan writers who make no mention of Christianity in their works, some of which were voluminous; and there are a number of Christian authors who do not refer to their religion when writing on pagan subjects. This conventional practice was once characterized as a "conspiracy of silence." Lastly, no deductions about Solinus' birthplace are to be drawn from the amount of attention he devotes to certain provinces of the empire. His allocation of space to localities was largely determined by Pliny, who, like other classical Roman geographers, was shaping his materials largely according to Greek patterns.

The title of Solinus' work, as indicated by the best evidence, was *Collectanea rerum memorabilium,* "Collection of Remarkable Facts," a title which provides a key to the book's popularity. Further clues are found in Solinus' prefatory remarks explaining that the book is a handy compendium designed to win approval of readers not by ornateness of style but by the scope and multiplicity of topics covered; that it comprehends the nature of men and animals, the strange manners and practices of exotic peoples, information about exotic trees, and so forth, all gleaned by laborious scholarship from the best authorities.

Solinus hit upon the secret of Pliny's popularity and produced a treatise that was more attractive to readers than the work from which it was largely derived. Adopting Pliny's geographical books as his framework, he drastically reduced Pliny's lists of place names, retaining the more interesting names and associations, interspersing them with a few hundred livelier bits extracted from Pliny's nongeographical books—mainly from VII–XII and XXXVII, with scattered excerpts from twelve other books. He also took 38 excerpts from Mela's *Chorography*—according to Mommsen's count—relating mostly to peculiar manners and customs. As we have come to expect by now, Solinus says not a word about his debt to Pliny and Mela, though he cites a rather impressive array of 63 authorities, ranging from Homer and Anaximander to Zoroaster. It is worth noting that half of his alleged authorities are also listed by Pliny, and their names were undoubtedly obtained from Pliny; it is likely that nearly all the remaining citations were also drawn at second or third hand from earlier works. Solinus has been commended for being the first writer to mention the Isle of Thanet, to remark about the absence of snakes in Ireland, and to provide sundry other bits of information on the British Isles, as well as for being the first to use the designation Mediterranean for the inland seas. All of this is very interesting, but none of these innovations is to be credited to Solinus himself.

The investigation of Solinus' precise relations to his sources and borrowers has proved to be a fertile field of research for students of classical and medieval Latin science. The traces of borrowing are abundant and the lines of transmission, either direct or through presumed intermediaries, are clearly discernible. A comparison of the text of Solinus with parallel passages in the extant writings of Latin geographers before and after him provides us with an illuminating example of the sort of copying that was indulged in by popular authors of scientific works.

Most of the spadework in this field has already been done by an eminent scholar. Students of ancient science are indeed fortunate that the greatest of German classicists, Theodor Mommsen, saw fit to lavish some of his most expert scholarship on Solinus' trivial work. Mommsen's prodigious industry and scholarly output call to mind the proclivities of Pliny and Varro, but, unlike the ancient encyclopedists, Mommsen understood how to synthesize and interpret his profusion of raw data. By tracking down every statement of Solinus that was derived from Pliny and Mela and by citing specific passages, drawn directly or indirectly from Solinus, in the works of Ammianus Marcellinus, Augustine, Capella, Priscian, Isidore of Seville, Aldhelm, Bede, and Dicuil, Mommsen has provided an invaluable guide to the scholarly methods of the ancient encyclopedists. In addition to the 38 extracts from Mela he found approximately 1,150 from Pliny in a book of Solinus that numbers less than a hundred pages. Of the passages of unknown provenience Mommsen concludes that some must be attributed to Solinus' own ratiocinations, so trifling that we can be thankful they are few. This is not the appropriate place to discuss Mommsen's reasonable arguments, heatedly disputed by several scholars, that Solinus was not using Pliny's original work but rather a Plinian chorography, distilled from his geographical books with later additions by some compiler in the second century; or his theory that Solinus' historical data were derived from a chronicle by one Cornelius Bocchus. Merely as an illustration of Mommsen's cogent arguments, we might point to his observation that Solinus and Apuleius, in passages originating in the same passages in Pliny, use identical phrases that are not found in Pliny.[2]

It very quickly becomes clear that Solinus was not pillaging Pliny in a mechanical fashion. Pliny's phraseology is sometimes reproduced and sometimes altered or inverted; nominatives become accusatives or ablatives, active verbs become passive, and in groups of items Pliny's order is often changed. Solinus follows Pliny's order in dealing with countries and islands, but deftly inserts sidelights on men, customs, animals, trees, and gems, extracted from widely scattered books of Pliny's encyclopedia. He may have been thoroughly familiar with the contents of Pliny's work, but it is more likely he was depending upon the detailed indexes that Pliny provided his readers in Book I. Thus in putting together an account of a single locality he gathers data from several of Pliny's books.

Solinus also makes some bad blunders, misreading or misinterpreting Pliny's text. Pliny gives the distance from Delos (*ab Delo*) to Paros and elsewhere the distance from the Fut River to the Diris (*ad Dirim*) Mountains; out of the prepositional phrases Solinus creates a town of Abdelos and a Mt. Addiris. Pliny records that, of the nine mountains in Achaea, Scioessa is the best known; Solinus makes Scioessa a place of nine hills. Despite mistakes of this sort and deviations from Pliny's phraseology, Solinus' textual correspondences can be of real service, as Mommsen pointed out, to editors engaged in emending Pliny's text.

Solinus begins his survey of the Mediterranean regions, not, according to standard practice, at Gibraltar, but with an extended excursus on Rome—surely an appropriate and attractive opening for Latin readers.

The first eight pages of his *Collectanea* discuss the name of Rome, the myths of its prehistory and founding, the date of the founding, the early period of the kings, and make a very cursory survey of history to the time of Augustus—in all, one of his most valuable sections because the material is not paralleled by any existing writings. In beginning his Mediterranean periplus with Rome and Italy, Solinus may have been following a Varronian tradition. Varro wrote a second large encyclopedia, in forty-one books (also lost), *On Human and Divine Antiquities,* in which six books in the early part are devoted to places and regions. It is believed that Varro focused most of his attention upon Rome and Italy and gave cursory treatment to the rest of Europe and to Asia and Africa. The remainder of the first chapter of Solinus' *Collectanea* is a collection of bits of gynecological folklore and reported instances of monstrous births, drawn almost entirely from Pliny's seventh book. Chapters 2–6 deal with Italy and adjacent islands. Like Pliny, Solinus dwells at length upon islands, and devotes almost as much space to Corsica, Sardinia, and Sicily as to Italy. Islands, as isolated topographical features, provide variety from the monotony of extended stretches of continental coast line. We note in Solinus the beginning of a trend which becomes increasingly evident in the works of later geographical writers in the Plinian tradition, of reducing Pliny's tedious lists and giving greater attention to separate and conspicuous features. Solinus skips along lightly, picking out of Pliny's text places and islands that have alluring associations. The regional survey thus becomes a scanty framework upon which to tack Pliny's antique and curious lore.

Following Pliny's order in general, Solinus next takes up Greece, Thessaly, Macedonia, Thrace, and the Greek islands (chaps. 7–11); the Hellespont, Pontica, and Scythia (12–19); Germany (20), Gaul (21), Britain (22), and Spain (23). Pliny, it will be recalled, handled the Mediterranean and Atlantic sides of Spain and Gaul separately at the beginning and end of his periplus. Solinus, having begun with Italy, deferred these countries to the end of his northern periplus, then with commendable judgment combined his borrowing from both of Pliny's accounts.

Solinus' southern periplus, like Pliny's, begins at the Straits of Gibraltar and ranges along the North African coast to Egypt, in orderly fashion until he comes to Pliny's reports of goat-pans, headless men, strapfoots, and satyrs. Here, with obvious relish, he gathers the other fantastic stories about dark Africa that Pliny saves for the homeward part of his periplus, accounts of dog faces and dog milkers, men who have a dog for a king, one-eyed and four-eyed men—all put together in one extended orgy of monstrosities (chaps. 24–32). Then he proceeds, according to Pliny's order, to Arabia and Syria (33–37); Asia Minor (38–45); Assyria, India, and Parthia (46–56); ending with the Gorgades and Hesperides islands.

It would be hard to find a more striking example of the deterioration that took place in the transmission of a classical work to the Middle Ages than Pliny's geographical books. A distillation of these four large books, together with approximately six hundred extracts drawn from other books of Pliny's encyclopedia, provides the bulk of the matter contained in Solinus' treatise of under a hundred pages. Martianus Capella covers the same ground, a century or two later, in a book one-fourth as large, drawing his material from Solinus and Pliny—again without a word of acknowledgment, though he lists as his sources Dicearchus, Archimedes, Anaxagoras, Pytheas, Eratosthenes, Ptolemy, and Artemidorus, men who are nothing more than names to him. Capella, at any rate, must be commended for discrimination in selecting his professed authorities, if for nothing else. Isidore of Seville distilled several hundred items extracted from Solinus' *Collectanea* in the books on animals, cosmography, stones, and agriculture in his encyclopedia; but his section on regional geography is only a few pages in length. By his time —the seventh century—nearly all traces of the science of geography have disappeared. Isidore sweeps across continents, listing a few names of countries for each; but to the tiny Isle of Thanet he gives a separate paragraph in his chapter on the Islands of the World, because of Solinus' intriguing statement that there are no snakes on the island and that earth taken from it to any part of the world kills snakes. Etymology is the basis for Isidore's encyclopedia; he concludes that Thanet is so called because it is death (Gr. *thanatos*) to snakes. Practically all that is left of Pliny after five centuries is a few anecdotes and *incredibilia*.

Solinus gives abundant evidence of his inability to interpret and carelessness in handling Pliny's text; Capella and Isidore are much worse in their misinterpretations of Pliny and Solinus. It is not easy to distinguish between borrowings from Pliny and from Solinus because both authors were widely circulated in the Middle Ages. Augustine and Capella must have been consulting the texts of Pliny and Solinus side by side, because many of their borrowings appear to have been conflations from both writers.

Mommsen's tabulations of the passages extracted from Solinus' *Collectanea* by nearly a dozen medieval compilers provide the concrete evidence of Solinus' popularity in the Middle Ages. The 153 extant manuscripts that Mommsen lists are further evidence. By 1554 five printed editions had appeared and the *editio princeps* had been reprinted numerous times. Itself a drastic reduction of earlier sources, the *Collectanea* was in turn re-edited and epitomized by several medieval hands. In the sixth century it was revised and given the new title of *Polyhistor*, which through some confusion became the surname of the author in the minds of later scholars. It was abridged in hexameters in the tenth century by a certain Theodericus and again in the twelfth century by Petrus Diaconus. It continued to be widely read and to exert an influence into the Renaissance.

Difficult editorial problems await anyone who undertakes to prepare a translation of an inexpertly and drastically reduced compilation such as the *Collectanea*. If the translator faithfully adheres to the text, his version will strike his readers as strange and at times nonsensical. If he attempts to gloss over the author's carelessness and ignorance by means of a polished or ambiguous translation, he will be creating a false impression of the author. To indicate the precise relationship of the compilation to the text or texts the compiler was abstracting and interpreting, he has to devise elaborate apparatus and document heavily, and his translation becomes unreadable. In no case can a modern translator hope to gain that approbation and esteem that were accorded the Elizabethan translators of such works. Contemporary readers of Philemon Holland's version of Pliny (1601) and Arthur Golding's version of Solinus (1587)[3] supposed that they were gleaning reliable information about the wonders of the world. And there is of course an appropriateness which modern readers still appreciate about an Elizabethan rendition of accounts of Pliny's monsters.

Comparable problems exist for the editors of the Latin texts. U. F. Kopp remarks at one point in his 1836 edition of Martianus Capella that if one makes the necessary corrections in the reading of the text one will in fact not be emending the reading but correcting Capella's mistakes. Theodor Mommsen's edition of Solinus still serves as a model for textual critics. But the modern editors of Capella have been unable to make up their minds whether to offer a text based upon the manuscripts or an emended text which presumes either that the author himself was using faulty texts or that later scribes were responsible for the blemishes.

We come now to the first of three Latin Neoplatonic encyclopedists who, within a century, were to write works that would transmit Platonic cosmography to the Middle Ages and would place their authors among the leading authorities on classical science, especially during the eleventh, twelfth, and thirteenth centuries. These men were Chalcidius, Macrobius, and Martianus Capella.

Probably the most important single factor in promoting and maintaining interest in Platonism throughout the classical period was the enduring attraction that both Greek and Roman intellectuals found in reading treatises inspired by Plato's *Timaeus* or the commentaries that undertook to interpret it—those writings which form the backbone of the handbook tradition of Greek and Roman theoretical science. The ancients regarded the *Timaeus* as so comprehensive in scope and so abstruse in meaning that works which essayed to elucidate it appeared to them to be disclosing the inmost secrets of nature and propounding the system of the universe. That interest in this work continued unabated during the Middle Ages is evident from the popular reception accorded the writings of the Latin Neoplatonists. It is not surprising to find that in Raphael's painting of "The School of Athens" the book in Plato's hand is the *Timaeus*. Cicero made a free translation of it, but although a considerable part of this is still extant, it was Chalcidius and not Cicero who provided the Middle Ages with a direct contact with Plato's enigmatic but fascinating work. Chalcidius translated and wrote a commentary upon the first two-thirds of the *Timaeus*, apparently early in the fourth century. Since Plato's story of the lost continent of Atlantis was in this portion of the book, Chalcidius had a share in keeping this fiction alive.

Little is known about Chalcidius' life. Most scholars believe that he was a Christian and that the Osius or Hosius to whom he dedicated his work was the bishop of Cordova who was present at the Council of Nicaea in 325 and later became St. Osius. Others doubt or deny that Chalcidius was a Christian, pointing out that his fondness for displaying his skill in biblical exegesis and his familiarity with Judaeo-Christian controversy indicate a closer tie with traditions of Philo Judaeus than with Christian doctrines.[4] Whether he was Christian, Hebraist, or Neoplatonist—or as one scholar has suggested, a combination of all three—is hard to determine from internal evidence.

As we peruse Chalcidius' commentary, with its highly technical vocabulary and complicated matter, keyed to intricate diagrams, we might be led to suppose that the author was an adept scientist, far superior to the Latin writers of his time or of any period in antiquity or the early Middle Ages—perhaps a Greek theoretician writing in Latin. It was not until 1849 that the brilliant T. H. Martin [5] made the discovery that Chalcidius' lengthy discourse on planetary and stellar motions was nothing more than a free translation of the astronomical portion of the Platonic handbook of Theon of Smyrna. All Chalcidius can now be credited with, it seems, is an ability to grasp in substantial part some fairly involved material in Theon's *Manual*—though this in itself marks him as a man of distinctly higher mathematical competence than the Latin writers of his time. It is pointless here to go into the arguments— however much there is to commend them—that Chalcidius was not translating from Theon's work but from Theon's source, Adrastus.[6] The fact remains that we have the texts of Theon and Chalcidius to compare, and Adrastus' work has been lost.

Chalcidius generally adheres to Theon's text closely until he comes to material which he does not understand or feels is too complicated for his readers to follow; he either passes over the difficult discussion altogether, digests it, or substitutes a simpler explanation. Theon was scrupulous in acknowledging his indebtedness to Adrastus in many citations; but Chalcidius, following the Latin practice of concealing major sources, mentions neither Theon nor Adrastus among the forty-eight authorities he cites. When he found in the original work the phrase "as Adrastus says" or "as he [Adrastus] says," he omitted it, while translating or paraphrasing everything else. But among his authorities he cites such impossible or improbable sources as Orpheus, Musaeus, Linus, Thales, Pythagoras, and Timaeus. At least he did not claim Adam, Abraham, and Prometheus as his sources on astronomy, as some medieval writers did. We see in Chalcidius' citations clear indication of the fondness of Latin compilers for pretending to be drawing from venerable sources: all the authors cited antedate the Christian Era.

The exegetical portion of Chalcidius' work is a commentary in the ancient and not modern sense. It is over three hundred pages in length, nearly five times as long as his translation of the *Timaeus*. Sometimes a passage from Plato's dialogue serves as a thinly veiled pretext for embarking upon a discourse on some subject. Most of the time, however, Chalcidius relates his discussion to the dialogue, introducing excerpts from all the chapters of the middle third of the *Timaeus*, in consecutive order but passing over the first third almost completely. His practice is to quote extended passages, excerpting them without change from his translation, and then to comment upon them, or, more frequently, to use short excerpts—sentences or phrases—altering their structure and syntax to fit them into the context of his argument.

The opening chapters of Chalcidius' commentary deal with cryptic passages which attracted the interest of commentators from the time when the *Timaeus* first became the subject of intensive study and which frequently were the inspiration of separate treatises by Platonists. Chapter I takes up the creation of the universe and attempts to explain the obscure passage (*Timaeus* 31c–32c) in which Plato mingles Pythagorean arithmetic with Empedoclean doctrines of the four elements.

Chalcidius introduces four geometrical diagrams: first, plane figures, to demonstrate that one mean is sufficient to "bind" the extremes in a plane number; then parallelepipeds, to demonstrate that two means must be inserted between the extremes of solid numbers. In either proportion there must be equality of ratio to preserve the unity and to make the system perfect. In a corporeal universe fire is needed to make bodies visible and earth to make them tangible. Air and water serve as the means to bind the whole together. Parallelepipeds, with unequal sides, will best represent the four elements, since their qualities vary in intensity and amount. In celestial fire the most abundant quality is brightness, heat is moderate and present to a lesser degree, and solidity is the least common quality; in earth the most prevalent quality is solidity, then moisture in a moderate amount, and lastly light to a very small degree. It is here that examples of parallelepipeds are introduced, with sides of varying lengths, or, according to Pythagorean concepts, of varying units or numbers. The products of the numbers of the four parallelepipeds used are 24, 48, 96, and 192. To produce a sensible universe with perfect symmetry, the numbers employed represent a continued proportion.

Chapter 2 deals with an equally celebrated passage in the *Timaeus*

(34c–35b), on the origin and constitution of the World-Soul, probably the most puzzling passage in a dialogue notorious for its many enigmas. Chalcidius' explanation of the Demiurge's blending of Indivisible and Divisible, Same and Different in the creation of the World-Soul is one of the few extended discussions of that subject to survive from antiquity, but it throws little if any light on Plato's meaning. When Chalcidius comes to the portions that went into the fabrication of the World-Soul— the second portion double the first, the third one and one-half times the second, and so on—he arranges the numbers in a so-called "lambda diagram" with 1 at the apex, 2, 4, 8 on one side, and 3, 9, 27 on the other, a diagrammatic scheme that was used by Crantor,[7] the first in the long line of commentators on the *Timaeus*. He notes that in the World-Soul there are seven portions, a number which is commended by Pythagoreans as the best, most natural, and most sustaining of all the numbers in the decad. His discourse on the virtues of seven, its associations with the ages of man, quarters of the moon, parts of the human body, etc., is all drawn from Pythagorean sources.

Chapters 3 and 4 deal with the numerical ratios of the harmonic intervals in the musical scale used in the construction of the World-Soul (*Timaeus* 36a–c). A tetrachord consists of two major tones (ratio 9:8) and a Pythagorean semitone or limma (ratio 256:243). The octave consists of two such tetrachords which are disjoined, so that there is an interval of a tone between the last note of one tetrachord and the first of the other. Thus, in an octave there is a sequence of tone, tone, limma, tone, tone, tone, limma. Since the ratio 9:8 has no rational square root, it is not possible to divide a tone into two equal semitones.

There follows a lengthy excursus, in Chapter 5, on stellar and planetary motions, the section which Martin discovered was a free translation of the second part, almost a half, of Theon's handbook. The ostensible occasion for this extended account is a passage in the *Timaeus* (37a–c) in which the World-Soul discourses with itself about intelligibles and sensibles, and another passage (36b–d) about the circle of the Same and the circle of the Other. Chalcidius also quotes at the beginning of this chapter, in his own translation, a venerated passage from Plato's *Phaedrus* (245c–246a) about the immortality of the soul and the beginning of motion being self-moved. Since the last passage also appears twice in Cicero's works (*Tusculan Disputations* I and the *Dream of Scipio*) and once in Macrobius' *Commentary on the Dream of Scipio*, and since all four writings are supposed to have been derived to a large extent, directly or indirectly, from Posidonius, some scholars have felt that Posidonius must have used the passage from the *Phaedrus* in his lost commentary on the *Timaeus*. Instead of addressing his attention to Plato's quoted passages, Chalcidius introduces here what amounts to a concise treatise on astronomy, all of it translated closely or loosely from Theon and adhering to Theon's order of topics throughout. Only after he has completed his translation of Theon's treatise does he discuss the Platonic passages in a few pages at the end of the chapter.

The account of astronomy begins with the conventional demonstrations of the sphericity of the universe and the earth, all rather loosely translated from the opening paragraphs of Theon's section on astronomy. Hereupon Chalcidius becomes almost literal in his translation of Theon as he offers demonstrations, geometrical and real, of the spherical shape of the seas and lands. Numerous commonplaces, found in most of the handbooks, follow: that the earth is at the exact center of the universe, that the earth is a mere point in comparison with the size of the universe, and that the celestial sphere rotates on an axis drawn through the center of the earth. Next come definitions of the celestial circles, including arctic, antarctic, equator, tropics, zodiac, horizon, and meridian. Theon and Chalicidius keep the courses of the sun, moon, and all the planets within the zodiac. The fixed stars are borne along in the uniform rotation of the celestial sphere and never change their respective positions, points of rising, apparent size, or color. The sun, moon, and five planets, on the other hand, have independent motions in latitude and longitude, and are seen to change their apparent size and velocity as they range nearer to or farther from the earth. John Scot Erigena, in the ninth century, attributes to Plato an observation that the planets take on different colors according to the qualities of the regions they are traversing;[8] but it is likely that he was inspired by Chalcidius' remarks.

Motions in latitude vary for the different planets. Like most handbook authors, Theon and Chalcidius mistakenly suppose that the sun is inclined to the ecliptic, according to them an inclination of $\frac{1}{2}°$ that occurs in Libra. This faulty observation originated in the time of Eudoxus or perhaps before, and was corrected by Hipparchus, who demonstrated that the sun's apparent course does not deviate from the ecliptic. Some scholars accordingly were led to conclude that Theon's astronomy was pre-Hipparchan; but it would be unwarranted to draw any general conclusions about the dating of an agglomerate body of handbook traditions. Elsewhere Theon and Chalcidius introduce Hipparchan views and calculations. Saturn, according to both writers, has an inclination of 3°, Mars and Jupiter 5°, Mercury 8°, and Venus and the moon 12°. These figures coincide with or closely approximate the estimates of most handbook authors. The moon completes its orbit in 27⅓ days, the sun in slightly less than 365¼ days; Mercury and Venus have varying periods of more or less than a year, depending on whether, with quickened pace, they are catching up with the sun or, with slackened pace, are being overtaken by it. Mars' orbit is completed in less than 2 years, Jupiter's in approximately 12 years, and Saturn's in a little less than 30 years. The superior planets range freely away from the sun and are sometimes found diametrically opposite; but Mercury and Venus always stay close to the sun, Mercury's maximum elongation being 20° and Venus' 50°. Then follow the definitions of the heliacal and acronychal risings of the planets and stars.

Although Chalcidius later discusses Venus' revolutions about the sun, and presumably assumes similar motions for Mercury, he here introduces a discussion about the fixed order of the planets, a regular feature in the handbooks. Adhering to handbook traditions was much more important than consistency. He points out that certain Pythagoreans maintain the order moon, Mercury, Venus, sun, Mars, Jupiter, and Saturn, but that Eratosthenes placed the sun immediately above the moon. Chalcidius' statement, derived from Theon, is considered reliable testimony that Eratosthenes did approve of the old Platonic arrangement of the moon and sun in the first and second positions. Stoic doctrines about the sun being the heart of the universe probably had a lot to do with the prevalence of the later order, which located the sun in middle position among the planets. The Pythagoreans that Chalcidius refers to would then be Neopythagoreans.

The sun and moon have constant and direct motions into the signs behind them; the other planets, however, experience progressions, stations, and retrogradations. This strange behavior of the planets is apparent and not actual. The sublunary expanse is a realm of change, of birth, death, increase, decrease, and movement from place to place.

The underlying cause of all this change is the errant motions in latitude of the planets, particularly of the sun and moon, and the irregularities in the orbits of the other planets.

Chalcidius then takes up the anomalies in the quadrants of the sun's orbit. Of the total of $365\frac{1}{4}$ days, $94\frac{1}{2}$ are consumed in the sun's passing from vernal equinox to summer solstice, $92\frac{1}{2}$ from summer solstice to autumnal equinox, $88\frac{1}{8}$ from autumnal equinox to winter solstice, and $90\frac{1}{8}$ from winter solstice to vernal equinox. (These figures of Theon and Chalcidius coincide with those found in Geminus' *Introduction to the Phaenomena*, another vestige of Posidonian influence.) The sun's velocity appears to be slowest in Gemini and swiftest in Sagittarius. Chalcidius uses geometrical diagrams to explain the anomalies of the sun's orbit. Either an eccentric orbit or a scheme of epicyclic motions will account for the progressions, stations, and retrogradations of the planets. Again a diagram is introduced to assist the reader. Mathematicians prefer to have the planets describe clockwise epicyclic motions while the centers of the epicycles move along deferents in a counterclockwise direction. Natural philosophers prefer to have both the epicycles and deferents moving in a counterclockwise direction. Chalcidius erroneously supposes that these arrangements account for the stations and retrogradations equally well.

Chalcidius next discusses eclipses, using diagrams to show what forms the shadow assumes if the luminous body is larger or smaller than, or of the same size as, the occulting body. The sun and moon do not suffer eclipses every month because of the sun's deviation of $\frac{1}{2}°$ from the ecliptic in Libra and because the moon, according to Hipparchus, deviates $10°$ (Chalcidius' earlier figure one recalls was $12°$). A work by Hipparchus on the distances of the sun and moon is cited as authority for the statement that the sun is 1,880 times as large as the earth. Comparison of texts shows that Chalcidius got his citations of Hipparchus from Theon, but he omits Theon's statement that Hipparchus also estimated the earth's size as 27 times larger than the moon's. (Both figures refer to volume.) The sun's radius would be $12\frac{1}{3}$ times as great as the earth's and the earth's 3 times as great as the moon's. Consequently, according to Theon and Chalcidius, the shadow cast by the occulting body will be conical-shaped and partial eclipses will frequently occur.

In passing it might be observed that Theon did not consult Hipparchus' works to obtain his data either. The results of Hipparchus' observations and calculations had been incorporated in garbled form in the handbook tradition. It is unfortunate that for some of Hipparchus' most important work we have nothing more than the conflicting reports and discrepant data of compilers. Cleomedes' handbook, for instance, gives Hipparchus' estimate of the sun's size as 1,050 times greater than the earth's.

The discussion of eclipses concludes Chalcidius' translation and digest of Theon's astronomical section, a fairly faithful translation of the more elementary parts and a very cursory handling of the more difficult parts. One important astronomical passage in Plato's *Timaeus* still remains to be explained: the puzzling statement (38d) about the "contrary tendency" in the motions of Venus and Mercury. Chalcidius points out that some authorities suppose Plato's expression refers to the contrariness between the sun's natural motion from east to west and its epicyclic motion in the course of a year from west to east; others think it refers to the motions of Venus and Mercury, which alternately overtake or are overtaken by the sun and appear ahead of it as morning stars or behind it as evening stars.

Chalcidius then proceeds to explain the theory of Heraclides of Pontus, referring in the process only to the motions of Venus, although it is generally supposed he was implying that Mercury's motions were also included in the explanation. He imputes to Heraclides a scheme in which the sun and Venus have concentric epicycles, Venus appearing at times above and at times below the sun and never exceeding 50° of eastward or westward elongation. The evidence is clear that Chalcidius and Theon are mistaken in attributing an epicyclic system to Heraclides. His hypothesis was a simple one of heliocentric motions for Venus and Mercury, the theory of a natural philosopher and not of a mathematician. A moment later Chalcidius commits the more grievous error of ascribing a system of epicycles for the sun, Mercury, and Venus to Plato—who lived long before epicycles were devised by Greek mathematicians. In this case the epicycles are not concentric but separate, the deferents of Mercury and Venus being beyond that of the sun. Heraclides' theory of the heliocentric motions of Venus and Mercury is described or alluded to by a number of classical authors; but Chalcidius, even though he incorrectly represents Heraclides' theory, is one of the two ancient writers to attribute the system to him by name. The other is Simplicius.

In conclusion it should be pointed out that it is an extremely fortunate chance that has preserved for us two handbooks in a direct line of transmission, so that by comparing texts we have an opportunity to observe the sort of extensive borrowing in which ancient compilers regularly engaged. All of Chalcidius' theoretical science was derived from Theon or from his source; in the latter case Theon himself would have to be regarded as little better than a wholesale copier. We may look upon Chalcidius as a mere translator or derivative author, but the fact remains that because of his ability to comprehend a considerable amount of Greek theory he must be ranked very high among Latin scientific writers; he was probably the closest approximation to a Greek theoretician that the Latin world produced during the classical and postclassical age. He enjoyed a tremendous vogue during the Middle Ages,[9] particularly among Christian authors and the Chartres Schoolmen; and his commentary became, in turn, the subject of a commentary in the twelfth century. Portions of it, doubtfully attributed to William of Conches, have survived.

X

FIFTH-CENTURY NEOPLATONIC COMMENTATOR

Macrobius and Martianus Capella, the two men who besides Chalcidius transmitted Platonic cosmography to the Middle Ages, both flourished early in the fifth century. In their writings we can see the Dark Ages closing fast upon the Western world. A sense of the decline of classical culture and a feeling of despondency over its waning are discerned in the minds of intellectuals like Macrobius who, as the empire was collapsing in the West, were fervently clinging to the remnants of Roman learning and glory. The schooling of non-Christians and Christians alike was secular; and even Christian zealots had a share in perpetuating classical traditions through their books. The elite of statesmen and intellectuals had to join hands to foster and preserve the vestiges and memories of Rome's achievements.

A realistic performance of latter-day dilettantes displaying their classical erudition is enacted for us in the seven books of Macrobius' *Saturnalia*, the last classical representative of the popular genre of banquet literature. The models for this were Plato's *Symposium* and Xenophon's *Symposium*; important examples are the *Symposiacs* and *Banquet Questions* of Plutarch and the *Banquet Sophists* of Athenaeus, all five of which have survived. In Macrobius' work a number of distinguished men have been invited by Vettius Praetextatus, eminent statesman and functionary of pagan religion, to engage in conversation and debate during the holidays of the Saturnalia. Among the invited guests are Aurelius Symmachus, a high public official and author of a collection of ten books of letters, matching that of Pliny the Younger; the poet Avienus; the erudite Servius, grammarian and commentator on Virgil's works; the brothers Caecina and Furius Albinus, the one a prominent statesman, the other a scholar, author of treatises on dialectic, geometry, and music; the fabulist Avianus, whose collection of forty-two fables, written in elegiac meter and in the style of Babrius, was dedicated to Macrobius; Nicomachus Flavianus, ranking official and historian, author of a book of *Annals* dedicated to Theodosius; and the learned philosophers Eustathius and Horus—truly a distinguished company, including some of the best minds and most eminent men of affairs of that time. Praetextatus, Symmachus, and Flavianus were leading opponents of Christianity, but one of the Albinus brothers was probably a Christian and the other had a Christian wife. Macrobius does not introduce himself into the dramatic setting of this banquet, but is believed to have represented himself in the person of one of the guests, Evangelus. In any case, these banqueters are kindred spirits of Macrobius, who himself combined the leisurely pursuits of learning with a devoted and distinguished career of public service. Several of the guests in the narrative were friends of his in real life.

In the remarks addressed to his son with which he opens the book, Macrobius wonders whether the Cottas, Laeliuses, and Scipios of Republican Rome will continue to serve as exemplars of gentility and learning in Latin literature, and reveals his anxiety lest the society of his own day will be unable to offer anyone to match them. Surely the endowments and accomplishments of the banquet guests are on a par with those of luminaries of classical Rome. The fervency of Macrobius' words and the ingenuousness of his attitude suggest that he believed the men of his day were the equals of the notables of Roman history; but he immediately betrays his own sense of insecurity and inferiority by plunging into the compilation of a lengthy tome culled substantially, and with fraudulent concealment, from earlier sources.

The scene is the home of Praetextatus on the eve of the Saturnalia. Each day during the holiday symposium is to be devoted to serious and learned discussion and debate, but at dinner time the mood is to change completely and nothing but pleasant conversation and jovial banter are in order. Typical of the discussions to ensue on succeeding days, the conversation opens with a momentous question: Exactly when does the Saturnalia begin, or, to put it in another way, when does tomorrow's day begin? Punctilious considerations about the calendar and feast days had always intrigued Roman antiquarians. Caecina Albinus is the first to answer. He recalls reading what Varro had to say on the subject in his *Human Antiquities*, and apparently quotes a lengthy passage from memory. His fellow guests compliment him on his remarkable powers of recollection; but the first two pages of Varro's explanation were actually copied verbatim, with a few slight changes, from Aulus Gellius' *Attic Nights*. Thus Macrobius proceeds, putting into the mouths of the banqueters extended discourses that he has taken from Gellius, from Plutarch's *Banquet Questions*, or from other works. It is surmised that he was using an earlier version of the *Banquet Questions* than the one we possess. The greater part of his pilferings from other authors will remain untraced because most of the works from which he drew have been lost; but it is reasonable to conclude from the character of the volume that there is very little if anything of substance in it that is not compilation.

This heterogeneous collection of classical lore and antiquities has survived in large part; the few portions missing include the second half of Book IV. Modern specialists in many fields of Roman antiquities have found the *Saturnalia* a mine of information, valuable for the unique scraps it preserves from lost sources. It is a particularly important work to students of ancient religion. To the general reader it is another tedious collection of the sort that literary historians have come to associate with Latin writers aspiring to reputations as erudites and antiquarians. The most valuable portion, Books III–VI, is a critique of Virgil's poetry; but, sad to relate, Virgil by now has become just another man of learning, an infallible and omniscient authority. There is not the slightest suggestion of appreciation of his poetic qualities. This was to be the attitude of the Middle Ages toward Virgil down to the time of Dante.

Before turning to the work of Macrobius that is of interest to students of Roman science, let us consider for a moment what is known about his life. It is reasonable to identify the author Macrobius with the official who held high governmental positions in 399/400, 410, and 422, as recorded in the Theodosian Codex, and to assume that his vernacular was Latin and not Greek. The objection has been raised that Macrobius would have had to be a Christian to hold such high offices during the reign of Honorius and that he makes no mention of Christianity in his works.[1] Instead he writes with warm admiration in the *Saturnalia* about Praetextatus, Symmachus, and Flavianus, prominent opponents of Christianity at that time. The practice of withholding reference to Christianity in works on pagan subjects must be recalled, however, and the fact that even Christian authors and men of Christian sympathies, like Claudian, Synesius, Nonnus, and Boethius, when writing on pagan philosophy or learning never speak of Christianity. That Macrobius was a zealous supporter of the Neoplatonic school of philosophy is clear from his attitude in his *Commentary on the Dream of Scipio*; that he could also have been, at the same time, a nominal Christian who could qualify for holding important government offices is not at all unlikely. In the oldest manuscripts of his works he is given the designation *vir clarissimus et inlustris*, indicating very high rank in imperial service. The decree in the Theodosian Codex that records his elevation to the rank of grand chamberlain in 422 also calls him *vir inlustris*. With regard to his birthplace Macrobius merely says that he was born "under another sky," implying a great distance from Rome. Some scholars have thought his interest in Greek literature indicates he came from some Greek-speaking part of the empire; but, as Wissowa points out,[2] his numerous mistranslations of Greek words and phrases and his preference for Latin writers suggest a birthplace in some Latinized part of the Roman world, most likely in Africa.

The work of Macrobius that is of interest to historians of science is his *Commentary on the Dream of Scipio*, a volume whose title and avowed purpose are misleading; it is even less a commentary on that subject than Chalcidius' book was a commentary on the *Timaeus*. Macrobius took the closing episode of Cicero's *Republic*, corresponding to the Vision of Er at the end of Plato's *Republic*, and used it as a framework upon which to construct a treatise on Neoplatonic philosophy— the most satisfactory and widely read Latin compendium on Neoplatonism that existed during the Middle Ages. He shows not the slightest awareness of the anachronisms involved in imputing fully developed Neoplatonic dogmas to Cicero, who lived three centuries before Plotinus, the founder of Neoplatonism. In fact, Plato too is removed from considerations of chronology and becomes just another unassailable authority. One is reminded of a Renaissance painting or book illustration in which a host of masters, some living as much as six centuries apart, are gathered for a group portrait. Macrobius' *Commentary* is tiresome compared to the exquisite gem set in its title; it is seventeen times as long as the *Dream of Scipio*. However, because it is a lucid and compendious exposition of Neoplatonic doctrines and contains lengthy excursuses on such popular topics as dreams, Pythagorean number lore, cosmography, and world geography, it was a fascinating book to medieval readers. Most important of all, it was responsible for the preservation of Cicero's *Dream*.

A good example of the tenuous connection Macrobius' *Commentary* has with the *Dream of Scipio* is seen in the long digression on Pythagorean arithmetic and arithmology in chapters 5 and 6 of Book I. Scipio the Elder appears to Scipio the Younger in a dream and predicts that the latter will suffer death when his age is marked by seven times eight orbits of the sun, each of the numbers being full. This reference to Pythagorean climacterics gets Macrobius started on one of his longest digressions.

Full numbers are solid numbers, not in a physical but in a geometrical sense. Eight, being the first cubical number, is a solid or full number. Among its many attributes it marks the number of revolving spheres in the harmonious universe and it is the sum of two numbers that are neither begotten nor beget, namely, one and seven. (A number cannot be begotten if it is prime and it cannot beget if it does not produce a number within the sacred Pythagorean decad.) The number seven is full for obvious reasons. It has so many attributes and virtues that Macrobius devotes an entire chapter of eighteen pages to it. By now he has forsaken the *Dream of Scipio* and is purveying stock material found in the *Timaeus* commentaries, which, as we have seen, introduce Pythagorean arithmetic and number lore because of Plato's use of them in the *Timaeus* in discussing the creation of the universe, generation of the World-Soul, and harmony of the spheres. Macrobius comments on the same passages that Chalcidius used in his opening chapters. Some scholars assume that he was borrowing from Chalcidius' commentary,[3] but in that case there would have been verbal correspondences. Instead, nearly all the material in the last six pages of Chapter 6 is literally translated from the discussion of the number seven contained in pseudo-

Iamblichus' *Theology of Arithmetic*,[4] a work which is supposed to have been largely derived from Nicomachus' lost *Theology of Arithmetic*. It appears that Macrobius was using a Greek commentary on the *Timaeus* for his excursus on Pythagorean arithmetic, probably the lost commentary by Porphyry. Since the number seven is the sum of one and six, two and five, and three and four, Macrobius grasps the opportunity to discuss in this chapter all the numbers of the Pythagorean decad except nine.

Medieval scholars looked upon Macrobius as an outstanding expert on cosmography. His cosmographical section, occupying seventeen chapters (I.14.21–II.9)—nearly half of the *Commentary*—was frequently circulated as a separate manuscript or bound together with other excerpts and treatises on astronomy. Rubrics and marginal glosses in the manuscripts call attention to the beginning and end of his cosmographical excursus, and there are headings for the major topics.

Recent research inclines to the view that Macrobius derived his information on cosmography as well as on Pythagorean arithmetic mainly from Porphyry's lost commentary on the *Timaeus*;[5] and that view is in accord with the inferences of the present study. A comparison of Macrobius' material on cosmography with surviving works of a similar nature reveals that he was following Platonist-Neoplatonic traditions that presumably stem from Posidonius' lost commentary on the *Timaeus* and through him perhaps go back to Eratosthenes' lost *Platonicus* and in some topics even back to Crantor, the first commentator on the *Timaeus*. Macrobius avows, when differences of opinion occur, that he is upholding Plato; but he gives no evidence of having handled any of Plato's works. His cosmography is basically Platonic, but the details correspond with those found in the handbooks of Geminus, Cleomedes, and Theon, and in Pliny's section on mathematical geography. All four of these writers are commonly believed to be transmitting received opinions and data gathered by Posidonius. The Posidonian character of Cicero's *Dream of Scipio* made it easy for Macrobius, in his commentary on that brief work, to apply Posidonian handbook traditions to its interpretation.

Once again we find the conventional cosmographical materials of Hellenistic and Roman compilers—a spherical earth poised in the center of the universe and encircled by seven planetary spheres and an outermost celestial sphere, which rotates diurnally from east to west. The planets have their proper motions from west to east in addition to their more conspicuous motions from east to west; their east-west motions are the result of their being "dragged along" by the diurnal rotation of the celestial sphere. Macrobius uses round figures for the planets' periods, conforming approximately with those found in the handbooks dealing with astronomy. The planets' periods are determined by their distances from the earth: the more distant the planet, the longer its period.

Within his cosmographical section Macrobius makes four statements about planetary velocities that are hopelessly confused and irreconcilable. At one point he maintains that the planets travel at the same speed and that distance traversed is alone responsible for the difference in periods. But he had just represented the spheres of Venus and Mercury as encompassing the sphere of the sun; yet the three planets have periods of a "year more or less." Later he points out that his revered Platonists flatly rejected Archimedes' figures for planetary distances and decided that the correct estimate of the sun's distance is twice the moon's distance, Venus' distance thrice the sun's distance, and Mercury's distance four times Venus' distance. A moment later he explains that the high-pitched tones emitted by the outer planetary spheres are caused by the high speeds at which they revolve, and the deep tone of the lunar sphere results from its slow speed. He is quite unconcerned about consistency as he mingles doctrines originating in Pythagorean musical theory with later Pythagorean doctrines about planetary distances.

According to Macrobius the planets' eastward courses are confined within the belt of the zodiac. He is aware that their motions deviate from the ecliptic, but his account, which is less technical than those found in most handbooks, does not include degrees of inclination. The ecliptic bisects the zodiac and derives its name from the fact that an eclipse can occur only when the sun and moon are moving along the ecliptic at the same time. The sun can be eclipsed only on the thirtieth day of the moon, and the moon only on the fifteenth day. There is, to be sure, no mention of the regression of lunar nodes.

Macrobius' list of celestial circles is more comprehensive than most lists. In addition to the five celestial parallels he discusses the zodiac, ecliptic, Milky Way, colures, and horizon. He says that he does not deign to comment on the fabulous explanations of the Milky Way and instead is quoting the scientific opinions of Theophrastus, Diodorus, Democritus, and Posidonius. We may be sure that Macrobius did not go to original sources for these opinions, which are also found in the doxo-graphic literature. He probably got them from Porphyry: like Porphyry in *On the Cave of the Nymphs,* Macrobius assumes that the Milky Way crosses the zodiac at Capricorn and Cancer and not at its actual inter-section at Gemini and Sagittarius. Both Macrobius and Porphyry con-nect Homer's description of the cave at Ithaca with the Pythagorean portal of souls. Macrobius' inexpertness as a writer on astronomy is seen in his egregious blunders in stating that the colures are not believed to extend to the south pole and in substituting the visible horizon, some thirty-five miles in diameter according to him, for the celestial horizon in his list of celestial circles.

His using the form of a commentary on Cicero's *Dream of Scipio* to propound a system of Neoplatonic dogmas and his attitude that his two chief authorities, Plato and Cicero, are both infallible and omniscient were bound to cause Macrobius embarrassment at some time. His pre-dicament occurs when he comes to the determination of the correct order of the planets (I.19.1–10). This regular feature of handbook astronomy which, in other treatises, calls simply for a listing of the planets' posi-tions according to the duration of their orbits, becomes a complicated problem for Macrobius. Plato had adopted an early order which placed the sun immediately above the moon whereas Cicero was following the more popular later order, the one adopted by Geminus, Cleomedes, Vitruvius, and Pliny. The Ciceronian order was also, without doubt, that to which Posidonius gave his approval. It is clear that later Platon-ists were either taking an equivocal attitude, as in the case of Theon and Chalcidius, or abandoning Plato's order. Macrobius, however, was a blind worshiper of authority and did not have a sense of discrimination about technical matters. As a confirmed adherent of Neoplatonism, he felt obliged to prefer Plato's order and at the same time to offer an excuse for Cicero's mistaken view. In one place he presents Plato's order as moon, sun, Mercury, and Venus, and twice he gives it correctly as moon, sun, Venus, and Mercury. We are not to suppose that Macrobius had ever read Plato's works.

The confusion about the correct order, Macrobius asserts, is an ancient one: the Platonic order was originated by the Egyptians and the Cice-ronian one by the Chaldaeans—another example of the curious desire of compilers to make views more respected by ascribing them to age-old authorities. In Macrobius' mind there is not the slightest doubt about placing the orbits of Saturn, Jupiter, and Mars above the sun, because of their markedly longer periods, and the orbit of the moon beneath the sun, since its period is only twenty-eight days. The confusion arises over Mercury, Venus, and the sun, because their periods do not differ greatly. Macrobius says that the true order is disclosed by the fact that the moon is the only planet to shine by borrowed light; the other planets are located above the sun, in the realm of blazing ether, and consequently have their own light. Observers are deceived because there are times when Mercury and Venus are coursing through the lower tracts of their orbits and appear to be beneath the sun. Some authorities assigned an infrasolar order to those planets because they made their observations at a time when the planets appeared to be closer than the sun to the earth, and so Cicero's mistake is an understandable one.

Macrobius' ambiguous remarks about upper and lower courses of Venus and Mercury undoubtedly reflect something he had read but failed to comprehend about the heliocentric motions of those planets as first conceived by Heraclides. Historians of astronomy of the present century who have given attention to Macrobius—Dreyer, Heath, and Duhem—all mistakenly suppose that he is propounding the Heraclidean system here;[6] and if general histories of astronomy make any reference to Macrobius it is as one of the few ancients who adopted and transmitted the Heraclidean system. In the Middle Ages Macrobius' views were known as the Egyptian system and were interpreted as Heraclidean; Bartholomew the Englishman, Peter of Abano, and an anonymous court astrologer of Baudoin de Courtenay (1217–70) so regarded them; and William of Conches was patently confused by Macrobius' statements and says that, since the sun, Venus, and Mercury have nearly the same periods, their orbits are nearly equal in size and are not contained within each other but rather intersect each other. A careful reading of the first half of Macrobius' chapter (I.19) will serve to clear up these misapprehensions, early and late. This passage shows that Macrobius' statements cannot be taken as an exposition of the Heraclidean system, that he did not comprehend it, and that his vague remarks are even less clearly expressive of Heraclides' views than the statements made by Vitruvius about the motions of Venus and Mercury. Anyone who finds Heraclidean doctrines in Macrobius would have to be familiar with the system from the accounts given by other handbook authors.

The assumption of heliocentric motions for Mercury and Venus is of course incompatible with a fixed order of the planets, but this was no stumbling block to the compiler. Also incompatible with the fixed order are the Stoic doctrines, discussed earlier, about the sun, as the heart of the universe, controlling the movements of the other planets. Macrobius introduces these notions, sometimes referred to as "attraction-repulsion" doctrines, into his system as follows: "The sun is called the regulator of the other planets because it controls the departing and returning of each planet through a fixed allotment of space. Each planet has a definite limit which it reaches in its course away from the sun and then, as if forbidden to transgress, is seen to turn back; and again, when it reaches a certain point, it is recalled to its former direction. In this way the sun's power and influence direct the movements of the other planets over their appointed paths" (I.20.4–5).

The observation was made that a philosopher of the caliber of Posidonius, who is known to have adopted a fixed order for the planets and geoheliocentric motions for some of them, would not have been so inconsistent as to embrace solar "attraction-repulsion" doctrines at the same time.[7] But when it is pointed out that Cicero, Vitruvius, Pliny, Theon, Plutarch, and Chalcidius, as well as Macrobius, combined such irreconcilable sets of views in their accounts of the planets' motions and positions, and that all of these writers are believed to have drawn much of their astronomy directly or indirectly from Posidonius, it appears that he too must have had no compunctions about including them in one system.

Like a true novice Macrobius delights in amazing his readers with his brilliance as a scientist. While he is on the subject of the sun, he introduces two painstaking and elaborate demonstrations which are not paralleled in the handbook literature. The first has to do with the relative sizes of the earth and sun. He opens by expressing contempt for the methods and calculations of Eratosthenes and Posidonius, as he does later for the views of Aristotle and Archimedes. It is plain to see that he was far removed from these authors' works, which he knew only through the garbled or imprecise copying of generations of compilers. When Macrobius is in a mood to expose the errors of men like Eratosthenes and Aristotle, his representation of their views makes them appear to have been fools. Modern scholars who base conclusions upon statements that Macrobius and other late compilers make about scientists and philosophers of the classical age are likely to be misled just as readers in the Middle Ages were.

Macrobius pretends to be quoting from a work of Eratosthenes in which the sun is calculated to be twenty-seven times greater than the earth, and he goes on to say that Posidonius' estimate is "many, many times greater."[8] Their results, he points out, are based on lunar eclipses, and he chides them for using circular reasoning: they demonstrate that the sun is greater than the earth by proofs taken from lunar eclipses, and they account for lunar eclipses by assuming that the sun is greater than the earth. Macrobius is ready to solve the problem once and for all. He proceeds to estimate distances and orbits in millions of stades; and his procedure and figures probably astounded medieval readers much as recent estimates of intergalactic space in billions of light-years have astounded us. His alleged authority for this amazing demonstration is the ancient Egyptians.

The cone of the earth's shadow extends sixty earth-diameters, he baldly declares, without proof, and the apex of the cone just reaches the sun's circle. The figure happens to be approximately double the estimates of Hipparchus (according to Theon of Smyrna) and of Ptolemy for the distance of the moon; and Platonists commonly assumed that the sun's distance was twice that of the moon. The figure Macrobius used may have had this origin, but he would not have been aware of it. Using Eratosthenes' estimate of the earth's circumference as 252,000 stades—without crediting him, and although he had just pretended to be exposing Eratosthenes' errors in the very book in which the earth's circumference was calculated—Macrobius estimates the earth's diameter as 80,000 stades, "or slightly more." Multiplying by 60 he gets 4,800,000 stades for the distance from the earth to the sun's orbit. The diameter of the sun's orbit would then be 9,600,000 stades and, multiplying by $3\frac{1}{7}$, he calculates the orbit to be roughly 30,170,000 stades—a demonstration of involved computations that must have pleased him.

Next he measures the sun's shadow, cast upon the rim of a hemispherical bowl when the sun's orb is precisely resting upon the horizon, and gets an angular diameter for the sun of $\frac{1}{216}$ of its entire orbit. It is not surprising, considering the crude apparatus he describes for this ticklish operation, that his figure for the sun's apparent diameter (1° 40′) is grossly excessive when compared with the figures of Cleomedes (28′

48″), Aristarchus (30′), Ptolemy (33′ 20″), and Capella (36′), which approximate the actual mean figure of 31′ 59″. Dividing his figure for the sun's orbit by 216, Macrobius arrives at an estimate of slightly less than 140,000 stades for the sun's diameter.

He now points out that geometricians have shown that when the diameter of one sphere is twice as great as that of another, the first sphere is actually 8 times as great as the second. After all his tedious and, by his standards, careful computations, he is now satisfied to consider 140,000 as approximately twice 80,000, and so he concludes that the sun is 8 times as large as the earth. If he had taken the cube of $1\frac{3}{4}$, he would have gotten 5.359375 instead of 8, but it is unthinkable that Macrobius could have computed the cube of fractions. Had he been familiar with elementary geometry, he would have known that, if the earth's cone extended as far as the sun's orbit and the sun's orbit were circular, as he assumes it to be, the sun's diameter would then be twice as great as the earth's. Cleomedes (II.78) uses this geometrical proposition when discussing the sun's diameter. Macrobius, or his source, was evidently trying to impress his readers with elaborate computations and then had to strain to achieve the result obtained by simple geometry. Despite its crudity, Macrobius' estimate of the sun's size as 8 times the earth's was widely adopted in the Middle Ages.

Macrobius' second demonstration he also credits to the ancient Egyptians. The apparatus he describes was attributed to the Egyptians by Cleomedes, but Cleomedes and Capella used it to determine the angular diameter of the sun and moon. According to Macrobius, the Egyptians were able to mark off the twelve divisions of the zodiac by constructing two copper vessels, one having an opening at the bottom, like that of a clepsydra. First they placed the perforated vessel, filled with water, over the empty vessel. At the moment when a conspicuous star was rising above the eastern horizon, they removed the stopper from the upper vessel and allowed the water to flow into the vessel below. The flow continued for twenty-four hours until the moment on the following night when the same star returned to the horizon. The water that had run through was then divided into twelve equal parts. His ingenious Egyptians then procured two more vessels, each with a capacity of one of the twelve parts of water. They poured all twelve parts of water into the perforated container and placed one of the two small containers beneath it. As the constellation which they later called Aries first began to emerge on the eastern horizon, they let the water flow into the first container. Precisely as it became brimfull, the second container was substituted and notation was made of the conspicuous star on the horizon that was to mark the end of the sign of Aries and the beginning of Taurus. In this way they were able to record six demarcations in the belt of configurated stars along the ecliptic during one season, and the process was repeated six months later to mark off the other six divisions of the zodiac.

Like most handbook authors, Macrobius prefers to regard the numerical ratios that are found in early Pythagorean musical theory and in the blending of the constituent portions of Plato's World-Soul as determining the relative distances of the planets rather than to accept the estimates of professional astronomers. He peremptorily rejects Archimedes' figures and approves those found in Porphyry's commentary on the *Timaeus*. This is his only reference to the work which is usually regarded as his chief source. The sun, he says, is agreed to be twice the distance of the moon, Venus thrice the sun's distance, Mercury four times Venus' distance, Mars nine times Mercury's distance, Jupiter eight times Mars' distance, and Saturn twenty-seven times Jupiter's distance. Since these numbers correspond to the proportions of the World-Soul, and since Macrobius later shows that the ratios arising from them also produce the harmony of the spheres, it was natural for him to assume that these numbers also determine the distances of the planets.

In the concluding chapter of his section on astronomy (I.22), Macrobius presents his proofs that the earth is precisely at the center of the universe. He had shortly before rebuked Eratosthenes and Posidonius for resorting to circular reasoning. He now opines that "those reasons are truly incontrovertible which are mutually confirmed, the one substantiating the other and each arising from the other," and proceeds to give as fine an example of circular reasoning as one could hope to find: all things are drawn to the earth inasmuch as the earth, being the middle, does not move; it does not move because it is at the bottom (the center of a sphere is stationary and the middle is the bottom); finally, it must be at the bottom since all things are drawn to it.

He concludes by assigning conventional positions in the universe to the elements, a feature which was introduced into the handbook tradition by Aristotelians: the purest naturally has the highest position and is called ether; next comes the part which has slight impurity and

and weight, namely, air; then comes the part which is still clear but is corporeal, water; at the bottom are found the dregs, the solid matter of the universe, namely, earth. The earth is kept in position in the center of the universe by dense air that supports it on all sides.

Inept though he was as a writer on astronomy, Macrobius became an imposing figure in the traditions of medieval and Renaissance astronomy, and the crowning glory of his unwarranted fame came in his having a crater on the moon named after him. In fact the naming of craters is a good illustration of the impressive position held by the handbook astronomers in the traditions of the profession. Among the named craters are Firmicus, Cleomedes, Geminus, Posidonius, Pliny, Vitruvius, and Manilius.

Macrobius was no less highly regarded as an expert on geography than on astronomy during the Middle Ages. His geographical section (II.5–9) is very brief, yet is the clearest and fullest exposition of the intriguing global conceptions of Crates of Mallos to be found in Latin literature. It was destined to captivate the imaginations of medieval readers and to make Macrobius one of the most influential authorities in the field. Crates' conceptions, revived and made current by Macrobius and Martianus Capella, still held a dominant position in geographical thinking as late as the thirteenth and fourteenth centuries. Thus it is possible to trace a continuous connection between Homer's presumed geographical notions and views widely held in the late Middle Ages. Macrobius introduced the Cratean conceptions into his *Commentary* because he found echoes of them in Cicero's *Dream of Scipio*. Cicero is believed to have derived his Cratean notions from Posidonius, who, like Strabo and Eratosthenes, began his geographical treatise, *Regarding the Ocean*, with a historical treatment of Homer's views. Posidonius' geographical work was strongly influenced by Crates, who regarded Homer as an omniscient geographer and attributed to him advanced conceptions which could not have originated before Hellenistic times. It is reasonable to suppose that Macrobius derived his Cratean doctrines, like so much of the rest of his *Commentary*, from Porphyry's lost commentary on the *Timaeus*, a work which was in the Posidonian tradition. There are marked correspondences between details in Macrobius' geography and those found in the *Introduction to the Phaenomena* by Geminus, another Posidonian author who expounds Crates' system.

Crates, as we have seen, conceived of the globe as having four inhabited quarters, separated by an ocean girding the earth about the equator and another, branching off meridionally from the main equatorial ocean, girding it about the poles. Diametrically opposed to the known quarter would be the habitations of the antipodeans on a southwestern continent. The hypothetical inhabitants of a southeastern continent, referred to as antoecians, would be situated to the south of the known quarter, across the equator, and diametrically opposed to the perioecians, inhabiting a northwestern continent. These habitations would be eternally separated from each other because of the intense heat of the torrid zone and the impassable oceans.

Macrobius adopts Crates' views and carries the symmetry of his scheme farther than other ancient writers. He states with confidence that there is a sea corresponding to the Mediterranean in the temperate zone of the southeastern quarter, though he admits he is unable to verify the statement or to locate the sea on the map which was drawn to illustrate his discussion. Incidentally, the maps that accompanied manuscripts of his *Commentary* became models for one of the commonest types of medieval *mappae mundi*, the so-called "zone maps."[9] He also related the Cratean conceptions to a statement of Cicero's in the *Dream of Scipio*, pointing out how relatively small the Roman Empire is in a world which is largely uninhabitable and divided into quarters which are eternally separated. Boethius, reproducing Macrobius' views about the limited extent of the earth's habitations in his *Consolation of Philosophy* (II.7), attributes them to Ptolemy. Modern historians of geography acknowledge that speculations about other continents, stemming from Macrobius' doctrines, had some influence upon Columbus.

Wherever Cratean views were circulated in the Middle Ages, they were likely to stir up controversy. Not all intellectuals in classical antiquity could accept the idea of antipodeans. Lucretius and Plutarch openly scoffed at it. Among the Church Fathers, Lactantius and St. Augustine, by refusing to accept such notions,[10] served as guides to clergymen who came upon doctrines of the four habitations in the works of Macrobius and Capella. The possibility of three-fourths of the earth's inhabitants being permanently cut off from salvation was not a happy prospect to contemplate. Whether the clergy during the Dark Ages actually regarded the belief in a spherical earth and antipodeans as heretical has been a subject of heated controversy among modern scholars.[11]

Crates' explanation of the tides as the backwash from the branches of the ocean colliding at the poles was even in his day an extremely backward attitude. Scientists at that time were beginning to recognize a connection between lunar cycles and the tides. Posidonius had made careful studies of tidal phenomena at Cadiz; and when his findings were reported by Pliny and other writers, it appeared that the lunar explanation would prevail. But Macrobius, in becoming an authority on the subject, had a retrogressive effect upon the thinking of geographical writers for centuries. In the Middle Ages tides could be explained by two prevailing theories, the lunar or astrological theory, propounded by Posidonius, Pliny, and Bede, and the physical theory of Macrobius and Petrus Diaconus, or by a cautious combination of both sets of beliefs.

Still another dispute arose over the question of the habitability of the torrid zone. Proponents of the doctrines of Crates had to abandon his view that the tropics were the limit of human habitation, as explorers penetrated farther into the torrid zone, along the upper Nile or the coast of the Gulf of Aden. Eratosthenes had located the latitude of the Cinnamon Coast and Ceylon at 12°N, 8,400 stades south of the Tropic of Cancer. That Posidonius stoutly maintained the torrid zone was habitable we know because Cleomedes reports his arguments. Pliny, perhaps as a result of an expedition sent by Nero to trace the sources of the Nile, moved his southern limit of human habitation down to 4°N. But Macrobius, following handbook traditions that antedated the Roman explorations, gives the same dimensions for the zones as Geminus and Theon and admits habitations only as far as 4,600 stades south of the tropic. He assumes exact correspondence in the south temperate zone, with human habitations penetrating the borders of the torrid zone to the same extent as in the northern hemisphere.

According to Macrobius, natural philosophers have given the reason for the existence of an equatorial ocean: ethereal fire feeds upon moisture and so nature placed an ocean beneath the paths of the sun and other planets. This curious notion is found in the collection of doxographic writings and is repeated by Cleomedes and by Porphyry in his *Cave of the Nymphs*. Macrobius states that human habitations are found over the entire extent of the temperate zones. Inhabitants of the four quarters are diametrically or transversely opposed to each other but there is no fear of their falling off into space since they all have the earth, which is at the bottom of the universe, beneath them and the sky above. However, inhabitants of the northern and southern hemispheres do have opposite seasons, and the inhabitants of the eastern hemisphere have daylight when it is night in the western hemisphere.

Like Chalcidius and the Greek commentators on the *Timaeus*, Macrobius introduces a section (II.1–4) on the numerical ratios of the fundamental musical concords and an explanation of the harmony of the spheres. This time his pretext is a passage in the *Dream of Scipio* in which Scipio the Younger, who has been transported in a dream to the heavens above the corporeal universe, asks his adoptive grandfather to explain the pleasing concord of sounds that fills his ears. Scipio the Elder replies that the sounds are produced by the rapid motions of the heavenly spheres and that the concords are the result of the properly proportioned intervals of distance at which the planets revolve: the celestial sphere, the highest and swiftest, emits the highest note; and the lunar sphere, the lowest, emits the deepest note.

We have seen that the handbook scientists, awed by the weighty authority of Plato and Pythagoras, preferred to follow their doctrines about musical intervals determining the distances of planets rather than pay much attention to what Hellenistic astronomers had to say on the subject. They had an obvious cause of perplexity here in that there are nine spheres, if the earth is counted, and only seven notes in the octave. Most writers agreed that the earth, being stationary, would produce no tone; but they were still left with a problem, and there was little agreement among them about its solution. Cicero gives no indication of the embarrassment experienced by the Greek compilers. Scipio the Elder simply remarks that since two of the eight spheres move at the same speed, seven different tones are produced. But Cicero had just called Mercury and Venus "the sun's companions," and there would be three spheres moving at the same velocity. Macrobius draws attention to Cicero's remark and says that the three planets complete their revolutions in the same space of time, "a year more or less," but he ignores the discrepancy.

Macrobius opens his discussion of the numerical ratios in the concords with a story, reported by a dozen or more Platonist commentators and ancient compilers of handbooks on harmony, of how Pythagoras passed a blacksmith shop and noted that two smiths, striking a piece of metal in alternate succession, were producing sounds that were attuned to each other. He asked them to exchange hammers—and "the sounds followed the hammers and did not stay with the men." He then attached the hammers to gut chords and found that the same concords were produced but with a sweeter tone. These observations led Pythagoras to a study of weights and tensions of strings and to the discovery that the basic concords are produced by weights or tensions corresponding to simple numerical ratios. Pythagoras is credited by almost all modern authorities with discovering the numerical ratios in harmonious intervals, but it is assumed that his discovery was made by moving a bridge along a monochord and observing the relation between tones and the lengths of the chord. The story of the blacksmiths was obviously fabricated by Pythagorean admirers. Notes do not correspond to the weights of hammers, and the vibrations of strings do not vary as the weights stretching the strings but as the square roots of the weights.

According to Macrobius, Pythagoras discovered that the ratios producing harmony were few and simple: the sesquitertian (4:3) produces the fourth; the sesquialter (3:2) produces, the fifth; the double (4:2) produces the octave; the triple (3:1) produces the octave and fifth; the quadruple (4:1) produces the double octave; and the superoctave (9:8) produces a whole tone. The tone, by its very nature, cannot be divided into two equal semitones; so although the interval smaller than a tone is called a semitone, there is as little difference between it and a full tone as between the numbers 256 and 243. An interval of a fourth consists of two tones and a so-called semitone or limma; a fifth consists of three tones and a limma, an octave of six tones, an octave and fifth of nine tones and a limma, and the double octave contains twelve tones. This discussion about tones and ratios is stock information and corresponds with the material found in a score or more of manuals on musical theory and in Platonic commentaries. Macrobius points out here, as do other commentators, that the numerical ratios in the concords are the same as those used by the Creator of the World-Soul, and the World-Soul, in impelling the celestial spheres into motion, naturally instilled into them harmonious sounds originating in the fabric of its own composition. In the course of his discussion, Macrobius introduces statements of Plato from the *Timaeus*, about the numbers used in the construction of the World-Soul, and from the *Republic*, about the whirling motion of the spheres. Some curious remarks he makes correspond to statements found in Proclus' commentaries on the *Timaeus* and *Republic*; it is thought that both Proclus and Macrobius were using Porphyry's commentaries here.[12]

It is easy to see why Macrobius, handling such profound topics as the nature of the universe, the habitations of the world, and the harmony of the spheres in plausible and simple discussions, was one of the most attractive ancient writers on science to medieval readers. His *Commentary* was designed for lay readers, whereas Chalcidius' *Commentary* was for more serious students who would not turn away from several pages of unmitigated technical matter on one topic. The difference between the styles of the two writers is indicative of trends, as the ancient world declined into the Dark Ages: the most reasonable-sounding explanation or account, expressed in the simplest terms, was becoming the most desirable one.

XI

FIFTH-CENTURY VARRONIAN ENCYCLOPEDIST

Another giant step into the Dark Ages was taken by the third of the late Latin encyclopedists to be discussed here—Martianus Capella. This uninhibited compiler used the bizarre allegorical setting of a marriage of the god Mercury to a showily erudite young lady named Philology to introduce an encyclopedia of the seven liberal arts. The pendulum had taken almost a full swing. The nimble, versatile, graceful Hermes of the ancient Greeks being married off to this stilted personification of trite learning and handbook science! An account of events leading up to the wedding and a description of the elaborate ceremony, consummated in heaven, occupy the first two books. Then each of the liberal arts, in the person of a bridesmaid, is introduced in successive books, the order here being grammar, dialectic, rhetoric, geometry, arithmetic, astronomy, and music.

Capella handles each subject with complete self-assurance but with a better comprehension of the subjects of the trivium, since he was an advocate by profession, than of the mathematical quadrivium. He claims as his authorities many of the most conspicuous names in each field. That some of his authorities were mythological or of doubtful existence is indicative of his remoteness from acquaintance with his alleged sources. He cites Aristotle, Chrysippus, and Carneades for dialectic; Demosthenes for rhetoric; Euclid and Archimedes for geometry; Anaxagoras, Dicearchus, Eratosthenes, Pytheas, Archimedes, Ptolemy, and Artemidorus for geography; Pythagoras for arithmetic and astronomy; Plato, Eratosthenes, Archimedes, Hipparchus, and Ptolemy for astronomy; and Orpheus, Amphion, and Arion for music—about as ostentatious and meaningless a collection of names as if they were inscribed on the façade of a wax museum. At least they are spelled correctly, a virtue not always found in medieval writers. Among Capella's actual sources Varro seems to have been by far the most important. Close correspondences, including considerable verbatim copying, are found between portions of Capella's encyclopedia and the works of Aquila Romanus and Fortunatianus in rhetoric, between Capella and Pliny and Solinus in geography, and between Capella and Aristides Quintilianus in music. Again it must be pointed out that there were probably intermediaries who did the earlier copying, digesting, or translating of much of his borrowed material. His occasional use of Greek words and phrases is not to be taken as indication that Capella could handle the language well. He gives Greek names to allegorical figures and declines these names in Greek forms; he inserts a Greek proverb of four words in one passage; but his borrowings from Greek works appear to have been derived through Latin intermediaries, translations, or adaptations in handbook form.

The model for Capella's encyclopedia was Varro's *Nine Books of Disciplines*. As has been mentioned, one modern scholar, Ritschl, even attempted a reconstruction of Varro's lost work from Capella's *Marriage of Mercury and Philology*. Varro had added medicine and architecture to the traditional list of seven Greek liberal studies; but in Capella's allegorical setting, when it is suggested, near the opening of Book IX, that the bridal attendants Medicina and Architectonica lend a hand in the preparations, the suggestion is rejected; since they are concerned with mundane affairs, it seems appropriate to let them keep silence in the heavenly company. It is of no little interest that Varro's laudable attempt to introduce to Roman readers a full-scale offering of Greek liberal studies in one manual, even though it exerted a powerful influence upon Latin compilers in the separate disciplines, had to wait more than five centuries for an imitator. For most Romans, Varro was prodigiously learned but a little too academic.

Capella, in adapting Varro's encyclopedia to the tastes of readers of his day, finally established a canon of the seven liberal arts. There had not always been agreement about which subjects comprised the acceptable curriculum, some Christian authorities, for instance, rejecting astronomy because of its popular associations with astrology and other prophetic arts, and substituting philosophy. Capella's trivium and quadrivium became the foundation of medieval education. That he was successful in fastening upon the Middle Ages a completely academic program of secular studies, rejecting even Varro's practical arts of medicine and architecture, is a testimonial to the attractiveness of the allegorical form and style in which his book was written and also a witness to the ossification of intellectual and scholastic attitudes that was taking place in the fifth century.

There is little agreement about the details of Martianus Capella's life, but the most reliable evidence indicates that he was a lawyer, residing in Carthage, speaking Latin as his vernacular, and flourishing during the years 410–439. It seems reasonable to refer his remark in Book VI about the capital of the empire, *Roma armis viris sacrisque quam diu viguit caeliferis laudibus conferenda,* to the period after Alaric's sack of the city in 410—rather than to interpret it to mean that Rome was still flourishing, and on slim grounds to place Capella in the late third or early fourth century.[1] Other statements of the author make it appear that he was practising law at Carthage before Gaiseric's capture of the city in 439. Capella gives no evidence of Christian sympathies, and may have been a follower of the Neoplatonic school of philosophy.

Fantastic and bizarre as the setting and framework of the *Marriage of Philology and Mercury* are, the contents are no less extraordinary. As we scan its pages, we seem to be in a cuckoo land. We wonder how intellectuals of the fifth century could have derived pleasure and edification from reading this rehash, the bulk of which is feeble reflections of classical learning that was by his time five hundred to a thousand years old. That Capella was to continue to be one of the most popular authors for the next eight centuries makes the situation only stranger. We are confronted with the incongruity of an author, whose notoriously bombastic and difficult style has been attributed to his late date, barbarian origin, and extravagant attempts to outdo the florid style of his celebrated fellow townsman Apuleius, writing on the subject of rhetoric in

Book V as if he were a colleague of Cicero. Most of his examples and definitions are quotations or paraphrases from Cicero's speeches and his essay *De inventione*, derived as stock material from intermediate writers on rhetoric.

Following the Varronian order, Capella introduces his mathematical quadrivium with a book on geometry (Book VI). When the bridesmaid Geometria is finally presented, after a gaudy proemium consisting of three pages of hexameter and elegiac verses and some highly poetic prose, she opens her mouth not to discourse upon geometry, which is deferred for superficial treatment in ten pages at the end of the book, but to present in greatly reduced form the list of place names and geographical tidbits of Pliny and Solinus. Pliny's lists of thousands of place names were of little use to begin with, since he himself was almost devoid of geographical aptitudes; but at least Pliny gave an impression of diligence and concern for accuracy, as he recorded dimensions across provinces, occasional distances between places, and total distances across continents. When Solinus extracted juicy morsels from many of Pliny's books and inserted them in his condensation of Pliny's four geographical books to produce a distilled treatise of less than a hundred pages, which still retained some of Pliny's distances and his over-all dimensions, the result, as the title *Collectanea rerum memorabilium* indicates, was a book of wonders, arranged geographically. When Capella further reduced the geographical accounts of Pliny and Solinus to a section of approximately thirty-five pages under the head of geometry, which also preserved Pliny's over-all dimensions of terra cognita, the product was an absurdity. In his reductions of the earlier lists, Capella shows greatest fondness for place names that have literary associations.

He opens his geographical discussion in the conventional way with a section on mathematical geography, drawn almost entirely from Pliny through intermediaries.[2] There are the Plinian proofs of the spherical shape of the earth: celestial phenomena change with the latitude of the observer; there is a difference of hours when an eclipse is observed in the east and west; readings on sundials are seen to differ in latitudes 500 stades apart; there are 12⅔ (elsewhere 13) hours of daylight at Meroë at the summer solstice but at Ultima Thule six months of daylight. (Pliny had been unable to make up his mind about Thule, we recall, giving it 24 hours of daylight at the solstice in one place and six months of daylight in another. Capella chooses the less credible observation.) The equality of hours of day and night proves that the earth is located at the center of the universe.

Macrobius and Capella, as we have noted, were largely responsible for the currency of Cratean conceptions in the Middle Ages. Capella's chapters on the terrestrial zones contain a Cratean discussion of the habitations of the four quarters, not found in Pliny but the common property of handbook compilers. The temperate belts afford habitation over their entire extent. The peoples of the known world occupy the north temperate region of the upper hemisphere; inhabitants of the south temperate region across the torrid zone are called antoecians; those dwelling in the region diametrically opposed to Europe and Asia are called antipodeans; and the antipodeans of the antoecians are called antichthoneans. The seasons of Europeans and antoecians are opposite. At this point Capella becomes badly confused about Cratean doctrines. He says that Europeans and their antipodeans have common winters and summers and that only the antoecians have a view of the south pole, which is deeply hidden from inhabitants of the northern hemisphere. To inhabitants of equatorial regions all days and nights are equal in length and all celestial phenomena are visible. They experience two summers and winters each year as the sun passes by them, upward and downward, in its course between solstitial points. In polar regions phenomena are very different. The only celestial bodies having risings would be planets and these would not pass overhead but would be seen near the equatorial horizon for six months and then become invisible; the celestial equator would be on the horizon and only six signs of the zodiac would appear above the horizon.

Capella (at least in this place) correctly records Eratosthenes' measurement of the earth's circumference as 252,000 stades. He and Macrobius are credited by historians of geography with having transmitted this bit of information to medieval geographers.[3] Capella's account of the method used by Eratosthenes is also considered worthy of mention by historians; but it is actually so garbled that it is worthless. A correct account of Eratosthenes' procedure is found in Cleomedes' handbook. Eratosthenes, at Alexandria, measured the sun's shadow cast at the summer solstice in a hemispherical bowl and found it $\frac{1}{25}$ of a semicircle or $\frac{1}{50}$ of a great circle; he multiplied the estimated distance between Alexandria and Syene (directly beneath the tropic) by 50 to get his measurement of the earth's circumference. Capella reports that the shadow, multiplied 24 times, gave the measurement of a "double circle" and that the observations were made at Syene and Meroë at the equinox. Elsewhere (VII.858, 876) he incorrectly places Meroë beneath the Tropic of Cancer and Syene on the first climate north of it.

Capella next discusses the dimensions of the known world, purporting to be drawing his figures from Artemidorus and Isidorus (Pliny's alleged sources). He gives Artemidorus' figure for the length, from farther India to Cadiz, as 8,577 miles and Isidorus' figure as 9,818; and in attempting to reconcile the two he notes that Artemidorus added to his figure the distance from Cadiz to Cape Finisterre, 991 miles. All this is copied from Pliny through an intermediary; their figures do not always agree. Capella's estimate of the breadth of the known world, from the Ethiopian Ocean to the Don River, coincides with Pliny's figure (5,462); and he reproduces Pliny's list of stations used in computing that distance. Isidorus, according to Pliny, added 1,250 miles to that total, his estimate of the distance from the Don to the latitude of Ultima Thule; but Pliny regards this figure as a wild guess. Capella copies Pliny's figure and sagely repeats his skeptical attitude, with no mention of Pliny anywhere in his book. Pliny presents reasonable arguments for believing that Isidorus' estimate was too small; Capella ventures his own opinion that the length and breadth of the habitable world must be the same, for a globe cannot have sides of unequal length.

In accord with the regular practice for periplus geographers to offer stock proofs that the known world was surrounded by a continuous ocean, so as to make the coastal survey appear to be adequately encompassing all known countries, Capella repeats from Pliny the fabulous reports of Augustus (Pliny said Augustus' fleet) sailing past Jutland into the frozen Scythian Ocean, and of Macedonian sailors cruising along the shores of the Eastern Ocean from the Indian Ocean to the Caspian Gulf. He does not omit to report wrecks of Spanish ships in the Arabian Gulf and the circumnavigations from Arabia and Ethiopia to Spain and from India to Germany. The last report is an example of how badly an account of Pliny's can be garbled by the time it is recorded in Capella's book. Pliny says that the Swabian king gave Metellus Celer some Indian captives who had been driven off their course by storms; Capella says that Cornelius Nepos, after taking Indians captive, sailed through Germany. The author of an influential work on geography, Nepos somehow became associated with this dramatic adventure in later accounts.

Capella divides the known world into three continents, Europe, Asia, and Africa; from Pliny he adopts the Don River as the boundary between Europe and Asia, and the Nile as the boundary between Asia and Africa. He proudly demonstrates here, as elsewhere, that he is an eclectic scholar by remarking that very many authorities regard the Sea of Marmora as the boundary between Europe and Asia. His "very many" in this case is actually Solinus.

Pliny had set the pattern for medieval geographers of following the coastal survey form, which obliged the author either to omit interior regions or to give them scanty treatment. His train of copiers and digesters—Solinus, Capella, Isidore of Seville, and Dicuil, for example—all have enormous gaps in their internal geography. Dicuil, an Irish monk of the ninth century and the first geographer to write about the Frankish kingdom, reveals almost no geographical knowledge of that kingdom, and for his geography of Germany depends on Pliny, a writer whose account of that country is no more informative than those of very early writers like Herodotus. Medieval geographers followed the centuries-old practice of depending for their information on books rather than looking about them and checking their reading by their own observations.

The earlier part of Capella's geography was drawn, possibly indirectly, from Pliny; the later part from Solinus and Pliny. If he finds nothing noteworthy in Pliny's extended lists, he will omit as many as fifty sections without mentioning a single place name, covering his omissions by a summary remark: "Here are found many people, bays, rivers, and mountains." In the process of condensing the Plinian lists, Capella frequently joins place names that have no connection, and indicates that he has not grasped Pliny's meaning.[4] (Some of his mistakes, to be sure, are to be attributed to a faulty or illegible text.) Like a young student trying hurriedly to cull items for a term paper, he produces jumbled sentences consisting of elements found in scattered passages in Pliny's text. One typical example (VI.668) of his composite style will suffice: "Not far distant are seven mountains which, because of their equal height, are called 'The Brothers'; they are teeming with elephants and lie beyond the province of Tingitana, whose length is 170 miles."

In addition to his interest in place names with literary associations, Capella is always ready to pick up items of an incredible nature. He tries to enliven his southern periplus by following Solinus' precedent of highlighting the monstrous creatures of Africa and the marvels of India and Ceylon.

After completing his coastal survey of the southern continents, Capella finally gets to the subject of his book, his token treatment of geometry. This brief section consists of translations of most of the definitions of Book I of Euclid's *Elements*; a division of plane figures into rectilinear, curved, and mixed; some terminology applying to the construction of figures; the formal divisions of a proposition into enunciation, definition, construction, proof, and conclusion; a few Euclidean definitions dealing with proportional magnitudes, commensurables, and incommensurables; a list of the thirteen kinds of irrational straight lines; a few definitions of solid figures; and literal translations of the five postulates and first three axioms of Book I of Euclid's *Elements*. Euclid's opening definition, "A point is that which has no part," is translated "A point is that whose part is nothing." There is no reason to think Capella originated this blunder because it is quite clear he appropriated this entire section from some Latin Euclidean primer.

At the conclusion of the book the question is raised how one goes about constructing an equilateral triangle on a given straight line; and the heavenly company, immediately recognizing the problem as Euclid's first proposition, bursts out in a round of applause for Euclid. The bridesmaid Geometry then steps forward and presents a copy of the works of Euclid as a bridal offering—considering its rarity in the Latin world in those days, a suitable gift for a heavenly bride named Philology.[5] It is clear that Euclid's reputation was still quite alive in the fifth century, but any schoolboy who tried to learn geometry from Capella's textbook would get only a few introductory samples of his *Elements*. The study of geometry in the West remained in a deplorable state until the tenth century, with nothing more to rely upon than this scanty treatment in Capella's book and some Boethian digests.

Book VII of Capella's encyclopedia is devoted to arithmetic. In the greater interest that Capella evinces in this subject than in geometry, we have further tangible evidence of the effect that Nicomachus' *Introduction to Arithmetic* was having in the Latin West. It was because Capella was using Varro's encyclopedia as his model that he retained the old order of the quadrivium subjects, which gave first place to geometry.

After the usual poetic introduction, the bridesmaid Arithmetica begins her discourse with eight pages of familiar material on the virtues of each of the numbers in the Pythagorean decad, a segment of ancient pseudoscientific literature in which, as it has been pointed out, there are very close correspondences in the texts of more than a dozen extant writings.

Capella's treatment of arithmetic resembles that of Nicomachus' *Introduction to Arithmetic* and of Theon's section on arithmetic in his *Manual*, but his source is some intermediary among the many who widely circulated this material. He classifies numbers into even times even, odd times even, even times odd, and odd times odd, defines prime numbers, and then fits the numbers from 1 to 10 into those classifications. His first series of numbers is 1 to 9, the second 10 to 90, the third 100 to 900, and the fourth 1,000 to 9,000, "although some Greek writers appear to have added 10,000" (VII.745). He next discusses how square numbers are produced by the addition of odd numbers. Definitions of even and odd numbers follow, and of their subdivisions, and a classification of odd numbers into prime, composite, relatively prime, and relatively composite. Next come definitions of excessive, deficient, and perfect numbers; then a discussion of plane and solid numbers and the various kinds of plane numbers: triangular, square, and heteromecic. He discusses relative number—the relationships which numbers bear to other numbers; and relative quantity, taking up both equality and difference (the latter is not found in Nicomachus). Then come the Nicomachean ratios, with unfamiliar designations for superparticular and superpartient.

After his summary of Nicomachean arithmetic, Capella introduces material, near the close of his book, which is not found in Nicomachus but turns out to be garbled Euclid. Greek writers on arithmetic had appropriated such material from the arithmetical books of the *Elements*. Euclid's propositions about the product of an even number and an odd number being even, and that of two odd numbers being odd, Capella restates without proof. Other Euclidean propositions on the addition of a series of odd or even numbers he handles with numerical examples. He offers numerical treatment of many of the propositions contained in Book

VII of the *Elements*, as well as of the more elementary propositions of Books VIII and IX, but not in Euclid's order.

Book VIII contains Capella's treatise on astronomy, which despite its gross errors is the most orderly and comprehensive treatment of the subject by any ancient Latin author whose work is extant, and in these respects most closely approximates a Greek handbook. There are enough correspondences with Pliny's astronomy, particularly near the end of Capella's account, to indicate that both were derived from the same ultimate source. But although Pliny's discussion of astronomy is much more extended than Capella's, he shows his customary small interest in and comprehension of theoretical matters and instead gives way in many places to his penchant for folk beliefs and reports of miraculous occurrences. Capella on the other hand is academically detached and simply digests handbook materials. As a result we feel that Capella gives a good idea of the probable scope and contents of the sixth book of Varro's encyclopedia—an impression we do not get from reading Book II of Pliny's *Natural History*.

In order to encompass most or all aspects of a subject and still keep his encyclopedia within manageable limits, Capella has produced what amounts to a set of abridgments of the disciplines. The shortcomings of the quadrivium portion of his book result from this need for extreme conciseness and his inability to comprehend the mathematical subjects with which he deals. His conciseness and his florid style combine to make Capella one of the most difficult writers in all Latin literature. To be able to follow his discussion, one must be familiar with the subject matter of a discipline as it has been treated in other handbooks.

Like Macrobius, Capella speaks in reverent tones about the knowledge of the ancient Egyptians in astronomical matters. It has been thought that these two writers were merely confusing Pharaonic astronomers with Ptolemy or the renowned Greek astronomers who observed and worked at Alexandria in Hellenistic times.[6] More probably this reverence stems from a passage in Plato's *Timaeus* (22b), which was remarked upon by generations of commentators.

Capella's astronomical section opens with a conventional treatment of the celestial circles, defining first the five parallels, going latitudinally; next the colures, going longitudinally; then the oblique circles, and lastly the horizon. In his summary account of traditional materials the arctic circle rests upon the northern horizon and is always entirely visible; the tropics mark the northern and southern limits of the sun's divagations; the equator is equidistant from the tropics and marks the point where days and nights are of equal duration; the antarctic circle, at its upper limit, touches the southern horizon and is always entirely invisible. The two colures are great circles passing through the celestial poles, one originating in the eighth degree of Aries and the other in the eighth degree of Cancer. Capella claims Hipparchus as his authority for this information, but it was a regular practice in the handbooks to have the colures pass through the place where the solstitial and equinoctial points were located. The two oblique great circles, the zodiac and Milky Way, have breadth and are not mere mathematical lines. The zodiac, a belt 12° wide, stretches from one tropic to the other, bisecting the celestial equator, and confines the deviations of the sun, moon, and five planets. It has 12 divisions or signs, each a segment of 30° of arc. The sun (*sol*) is the only (*solus*) celestial body whose course follows the ecliptic. The Milky Way is the only visible great circle and is not recognized as one of the circles by some authorities. Capella oddly remarks that its circuit is much greater than that of the other great circles since it stretches from one arctic circle to the other and thus traverses almost the entire heavens. Its breadth varies, since it is seen to thin out from Cassiopeia to the sting of Scorpio. The last of the circles is the celestial horizon, separating the visible and invisible celestial hemispheres at any given time or location. Inasmuch as the horizon constantly changes with the rotation of the celestial sphere, it cannot be defined by constellations.

To assist in defining the location of the celestial circles Capella traces them through precise parts of constellations, e.g. the arctic circle from the head of Draco to the right foot of Hercules, through the breast of Cepheus to the front feet of Ursa Major, and back to the head of Draco. His tracing of the tropics and equator through the constellations corresponds almost exactly with the tracings found in Aratus' *Phaenomena*. Since Aratus was the great popular authority on the constellations in antiquity, we are not surprised to find material from his celebrated poem incorporated in the handbook tradition. When Capella discusses the antarctic circle, he writes like a medieval man. He proudly explains that he knows what constellations that circle enters, for no sector of the entire heavens is unfamiliar to him, but that he prefers to avoid giving the appearance of introducing false statements. There is no reason to suppose that Capella makes this claim because, as a North African, he had a view of the southern constellations. He is strictly a handbook writer, and in his day only constellations visible in the northern hemisphere were named and discussed. He then fixes the location of the colures by constellations, a feature which Aratus omits but which is found in Manilius' *Astronomicon*. However, there is very little correspondence between the constellations of Capella and Manilius, a writer who has frequently perplexed scholars because of the uniqueness of his material.

Capella also defines the locations of the celestial parallels mathematically, by dividing a meridian into 36 intervals and marking the intersections of the parallels along the meridian: the arctic circle 8 intervals from the pole; the Tropic of Cancer 6 intervals below that; and the equator 4 intervals below the tropic. The parallels of the southern hemisphere are at corresponding intervals. Macrobius uses different figures for these intervals: 6, 5, 4, 4, 5, 6. Thus Capella's intervals amount to 5° of arc and Macrobius' to 6°.

Following convention, Capella separates zodiacal constellations from those lying north or south of the zodiac. He prefers not to consider Capra, Haedi, Serpens, and Panthera as independent constellations but to identify them with the larger constellations Auriga, Ophiuchus, and Centaurus, which hold or support them. He lists 35 constellations outside the zodiac and concludes that it is not necessary to name the 12 zodiacal signs, since they are common knowledge. To help locate and identify the constellations, he records which constellations or their parts are rising when others are setting. Constellations rising vertically and setting transversely require more time to rise than to set. Cancer rises vertically in $2\frac{1}{12}$ hours and sets transversely in $1\frac{11}{12}$ hours. Then he gives the hours and fractions of hours required for the rising and setting of each zodiacal sign. When the sun enters signs that rise quickly, nights become longer, and when it enters signs that rise slowly, days become longer. And that explains why there is a difference in the duration of days and nights in the summer and winter despite the fact that there are always six zodiacal signs above and six below the horizon and the signs are all of equal dimensions, as proved by measurements with clepsydras. Geminus had raised the same perplexing question and had offered the same explanation, but in his account the equal spans of zodiacal signs had been proved by dioptra measurements. The close resemblances between

Capella's discussion and the accounts of Geminus in particular, and also of Cleomedes and Theon, make it evident that Capella was transmitting conventional materials.

Otto Neugebauer has very kindly drawn my attention to the startling fact that Capella at this point has recorded some astronomical data not found in any handbook traditions. His repository of worn-out and mutilated stereotypes is the last place in which one would expect to find unique observations, but such is the case. His figures for the rising times of the zodiacal constellations are in exact agreement with his statement a moment later that there are 14 hours 10 minutes of daylight and 9 hours and 50 minutes of darkness at the summer solstice. The perplexity arising from this statement is that the observer would have had to be located in a latitude slightly higher than that of Alexandria (Ptolemy's location), where the longest day is of 14 hours' duration, or lower than that of Rhodes (Hipparchus' and Posidonius' location), where the longest day lasts 14½ hours. There is one assumption we can make here with complete assurance: Capella did not time these observations himself, nor did he obtain the data from any North African astronomer. If he had gone to such trouble, it would be the only instance known to me of a Latin handbook author introducing an original scientific observation. Besides, Capella had no interest in precise timing. In one place he gives 13 daylight hours at Meroë at the solstice, in another place 12⅔ hours; he gives the same number of hours of daylight (14) and darkness (10) at Syene and Alexandria, though they actually differ by a half hour. Professor Neugebauer is inclined to suppose that the authority who was originally responsible for Capella's observations was making an interpolation of one-third of the difference between Alexandria and Rhodes. Who Capella's ultimate source was may never be known.

Capella raises still another question that he feels calls for an explanation: why does the sun, moving at uniform speed, require varying amounts of time to traverse different signs of the zodiac? He points out that although the earth is at the center of the universe, it is eccentric to the sun's orbit, the sun being sometimes borne at greater distance from the earth than at other times. As the sun is ascending the steeper tracts of its course, in Cancer and Gemini, it takes longer, lingering 32 days in Gemini; it spends less time in the lower tracts: 28 days in Sagittarius. Sojourns in the other signs vary between those extremes. Geminus raises the same question and explains it in the same way. Macrobius and Cleomedes also refer to the sun's steep ascent in Gemini as the cause of the retardation.

In appropriate order Capella next discusses planetary motions. He observes that the stars have fixed positions in the sky and so have only one motion, that of the diurnal rotation of the heavens. The planets also have this daily motion but in addition have contrary independent motions, each planet's period being in proportion to the distance it traverses in its own revolution in the reverse direction. The moon requires a month, the sun a year, and Saturn thirty years to complete orbits in a reverse direction over as much space as they traverse in one diurnal rotation of the celestial sphere. Their reverse motions are oblique to the direction of their daily motions. Capella points out that the Peripatetics have a different theory, maintaining that the planets are not really moving with a contrary motion but are unable to keep up with the speed of the celestial sphere and are being outdistanced, some rapidly, some slowly. It is a matter of indifference to Capella which view the reader prefers. Geminus had shown more expertness when he declared the position of "certain philosophers" to be untenable because the fixed stars describe parallel courses across the sky, and the planets, if they were merely lagging behind, would describe similar parallel courses. Instead, he says, the planets move obliquely, within the belt of the zodiac, some crossing from one side of it to the other. Moreover, stations, retrogradations, and direct motions of the planets indicate that they are moving independently.

Capella next notes that the independent motions of the seven errant bodies are alike in one respect, that they are all from west to east. The five planets experience stations and retrograde motions, but the sun and moon have only direct motions. The last two are eclipsed; the five planets do not suffer eclipses. (It is surprising that handbook authors seem to have been unaware of occultations of planets by the moon.) The three superior planets, like the sun and moon, have orbits about the earth, but Venus and Mercury do not; their orbits are about the sun. Capella is here clearly seen to be a proponent of the Heraclidean theory; and, because of the popularity of his encyclopedia, he became one of the leading transmitters of these geoheliocentric views which were to be so important in keeping heliocentric speculations alive in the Middle Ages. In his monumental work *De revolutionibus orbium caelestium* (I.10) Copernicus praises Capella's theory as an ingenious one.

The earth's position is eccentric to the orbits of all planets. In a universe that rotates with but one uniform motion no planet rises at the point where it rose the preceding day. The sun describes 183 parallel circles as it moves from the northern to the southern tropic, and describes the same circles in reverse order as it returns to the northern tropic. Mars describes twice as many parallel courses as the sun, Jupiter 12 times as many, and Saturn 28 times as many. Venus and Mercury also shift their daily risings and settings, but, as previously noted, their orbits are about the sun and not the earth. Capella foolishly remarks that they are sometimes borne in courses above the sun but are found most of the time below the sun. This erroneous notion seems to have arisen in the fact that when the two inferior planets are on the earth side of the sun they are more conspicuous. Capella observes that when both planets are below the sun Venus, with its wider orbit, is closer to the earth.

After his perceptive observations about the motions of Mercury and Venus, Capella reverts to the witless handbook practice of assigning fixed relative orders to the planets, in the very same paragraph in which Mercury and Venus are given heliocentric motions. He takes an equivocal position here, however, stating that some authorities place the orbits of Venus and Mercury above the moon, others above the sun; the orbits of Mars, Jupiter, and Saturn follow in that order.

After the general discussion of the planets, conventional practice calls for a brief statement about their relative distances, a matter that usually poses no problem for Latin writers since they are satisfied to let musical intervals determine the planets' distances. Capella, however, finds this an ideal opportunity to impress his readers with his mastery of the complexities of Greek astronomy, and he undertakes to measure the orbits of all the planets, something that not even Ptolemy attempted. His procedure faintly reflects methods used by Aristarchus, Hipparchus, Posidonius, and Ptolemy; but in the course of centuries handbook accounts had been carelessly transmitted and badly misunderstood, so that there are almost no correspondences between Capella's account and figures and the procedures and calculations of Hellenistic astronomers as reported in the Greek handbooks. Capella's calculations involve very rough and arbitrary estimates and are based upon a ridiculously precise and grossly exaggerated figure for the earth's circumference as 406,010 stades, which he attributes to Eratosthenes and Archimedes. (In Book VI, one recollects, he had recorded Eratosthenes' figure correctly as 252,00 stades.) Archimedes is known to have been interested in the earth's dimensions and to have preferred his own estimate of 300,000 to Aristotle's estimate of 400,000 stades. The Greek writers clearly indicate whether diameter or circumference measurements are involved and use a ratio of $3\frac{1}{7}$ for π. Capella seems completely unmindful of whether he is using diameter or circumference dimensions, and compares measurements of the earth's circumference and the moon's diameter (as a disc in eclipses or in measurements of its angular diameter) to get their relative sizes.

According to Capella the moon's dimensions may be determined by measuring its shadow on earth during solar eclipses. To illustrate, a total eclipse in the latitude of Meroë is seen as a partial one at Rhodes, and at the mouth of the Dnieper the sun is completely visible. From a knowledge of Rhodes' latitude and distance from the equator or from Meroë (which we have seen earlier he assumed to be beneath the tropic), it is calculated that the breadth of the moon's umbra is $\frac{1}{18}$ of the measurement of the earth's circumference. By measuring the distances of observation points north and south of the umbra, where the sun is partially eclipsed, it is estimated that the moon is three times as large as its umbra and consequently it is $\frac{1}{6}$ as large as the earth. (The details are omitted; Capella merely gives results.) The moon's angular diameter is determined by measuring and comparing the amount of water that runs through a clepsydra during the time it takes for the full orb to emerge on the eastern horizon and during one complete rotation of the celestial sphere. The moon's orb is found to occupy $\frac{1}{600}$ part of the complete circuit of the heavens (a fairly respectable figure of 36' for the moon's angular diameter). If the moon's apparent diameter is $\frac{1}{600}$ part of a great circle and the earth is 6 times larger than the moon, the moon's orbit is 100 times as large as the earth. Proceeding with his calculations and assuming that the planets have uniform velocities, Capella estimates from their periods of orbit that the sun's orbit is 12 times as large as the moon's orbit, and Mars' orbit 24 times, Jupiter's 144 times, and Saturn's orbit 336 times as great as the moon's orbit. If the moon's orbit is 100 times greater than the earth's circumference and Saturn's orbit is 336 times greater than the moon's orbit, it follows that Saturn's orbit will be 33,600 times greater than the earth's size. These impressive figures do not appear to have been widely adopted by medieval writers.

Next in order come the individual motions of the sun, moon, and planets. Half of the moon's sphere is always illuminated by the sun's rays, but because the illuminated side is turned away from the earth much of the time, the moon appears to be only partially illuminated. On the thirtieth day of its period the moon's bright side is wholly turned away from us and the planet becomes invisible. In its first reappearance each month it is crescent. At an eastward elongation of 90° it appears as a half-moon. At 135° it is at the gibbous stage and at 180° it becomes full. The same designations are applied in reverse to its corresponding phases as it is returning to its conjunction with the sun. In 24 hours the moon will advance 13° in the heavens, Mars $\frac{1}{2}$°, Jupiter $\frac{1}{12}$°, and Saturn $\frac{1}{28}$°.

The moon completes a revolution in 27$\frac{2}{3}$ days but requires 29$\frac{1}{2}$ days to overtake the sun because, during the time the moon is completing its orbit, the sun has moved into the sign behind and sometimes into a second sign behind. If the sun and moon have a conjunction in the last degree of Libra, Scorpio, or Sagittarius, the moon will overtake the sun not in the following sign but in the second sign behind. Thus the sun will sometimes pass through one of those signs unaccompanied; but, conversely, in the signs diametrically opposite, the moon will sometimes have two conjunctions. Inasmuch as it takes the sun 30 days to pass through some signs and 32 days to pass through Gemini, and since the moon catches up with the sun in 29$\frac{1}{2}$ days, two conjunctions will sometimes occur in these signs. Occasionally the moon reaches its full phase on the fourteenth, sometimes on the fifteenth, but most frequently on the sixteenth day of its period; but if a greater number of days elapses in the waxing stages, there will be a correspondingly smaller number of days in the waning, so that the total number of days in a cycle is always the same. A lunar year consists of 12 cycles of 354 days, 11 days less than in a solar year. The difference is made up in calendars by intercalations.

All planets keep their courses within the zodiac. Their deviations in latitude vary, ranging from 3° for Saturn to 12° for the moon and Venus. The latter two thus range over the entire breadth of the zodiac. The sun's supposed inclination to the ecliptic crops up here again, where it amounts to $\frac{1}{2}$° on either side and occurs only in Libra. The moon ranges from 6° north of the ecliptic to 6° south of it, crossing the ecliptic sometimes in sharp and sometimes in wide angles. The cycle of the regression of lunar nodes is completed in 235 lunations, or in the nineteenth year.

The moon is observed to return to the same position with respect to the celestial sphere on the same day of its period in 55-year cycles; and for all planets to return to identical positions with regard to the fixed stars a lapse of a great year is required. (Capella makes no estimate of the length of a great year, a matter of wild speculation among the ancient authors.) He then gives the Greek terms to designate the moon's course as it moves north or south of the ecliptic, away from or toward the ecliptic. If the moon happens to cross the ecliptic on the thirtieth day of its period, an eclipse of the sun occurs; if it crosses the ecliptic on the fifteenth day, the moon itself suffers an eclipse. Eclipses cannot recur within six months.

The daily shifting in the sun's points of rising causes it to describe 183 parallel circles (he again observes) as it ranges from the northern to the southern tropic; it retraces its course over the same circles in reverse order as it returns to the northern tropic. Although conditions in the northern and southern hemispheres correspond exactly, and although the distances of the tropics from the equator are the same, 185¼ days elapse in the sun's northward course to the Tropic of Cancer and 180 days in its southward course to the Tropic of Capricorn. This anomaly is due to the earth's being eccentric to the sun's orbit; while in the northern hemisphere of the heavens, the sun is elevated to the upper reaches of its course, but in the southern hemisphere it comes closer to the earth. To the inhabitants of the northern hemisphere the sun in Cancer brings summer, in Libra autumn, in Capricorn winter, and in Aries spring. Seasons are reversed for antipodeans.

Days in summer correspond exactly to nights in winter and summer nights to winter days; twice a year, at the equinox, days and nights are exactly equal. The amounts of daylight vary according to latitude. Capella recognizes eight latitudes or climates. His first climate is Meroë, which he here places beneath the Tropic of Cancer, although all other handbooks followed Eratosthenes in placing Syene on the tropic and Meroë several thousand stades south of it in the torrid zone. North of Meroë are the climates of Syene, Alexandria, Rhodes, Rome, Hellespont, Borysthenes, and a climate north of the Sea of Azov and south of the Rhipaean Mountains. Hours of daylight vary, depending upon climates, ranging from 13 equinoctial hours of daylight on the longest day and 11 on the shortest day at Meroë to 16 equinoctial hours of daylight on the longest day and 8 on the shortest in the climate of the Rhipaean

Mountains. Like Pliny, Capella is confused about arctic daylight. He observes that as one approaches the north pole he finds days getting longer and nights shorter. Consequently, directly beneath the pole there is perpetual daylight. Seasonal increases and decreases in amount of daylight follow a pattern: $\frac{1}{12}$ of the total amount of increase from winter to summer solstice is added in the first month, $\frac{1}{6}$ in the second month, $\frac{1}{4}$ in the third and the fourth months, $\frac{1}{6}$ in the fifth month, and $\frac{1}{12}$ in the sixth month. The explanation of this anomaly is that the zodiac bends around Cancer and Capricorn but cuts the equator almost directly.

Mercury and Venus, as noted earlier, have epicyclic motions and do not orbit about the earth. (Actually Capella's earlier reference to the motions of Venus and Mercury made the sun the center of the circular orbits and included no mention of epicycles.) Mercury's range of deviation in latitude is 8°. Since its maximum elongation from the sun is 22°, it cannot have acronychal risings. It does have risings and settings but appears above the horizon for only a short time. Mercury's last visibilities and visible risings recur in the fourth month, and not always at that time. Venus consumes over 300 days in its revolutions, ranges through 12° in latitude, and has a maximum elongation of 46°. At times it outspeeds the sun, and at times it falls behind; sometimes it travels above the sun and sometimes below it. It does not always complete its orbit within a solar year. If it is in retrograde motion at the end it takes longer than a year, and if it is in direct motion it sometimes completes its revolution in 11 months. When it is ahead of the sun it is called Lucifer, and when it is behind it is called Vesper. It is the only one of the five planets to cast a shadow. Venus' morning risings are frequently of 4 months' duration, its evening risings not more than 20 days. Its reappearances and last visibilities recur in periods of 9 to 10 months.

The three superior planets all experience risings, settings, stations, and retrogradations, but each has its own highest elevation, first station, and exaltation. Mars completes its orbit, eccentric to the earth, in approximately two years and has a deviation in latitude of 5°. Mars' apogee, that is, its point of highest elevation above the earth, is in Leo. Its first station is a unique one. Inasmuch as Mars' orbit lies closer to the sun than the orbits of the other superior planets, it feels the effect of the sun's rays even from a position of quadrature and undergoes a station 90° away from the sun on either side. Mars' exaltation or absis, its greatest distance from the center of its own orbit, occurs in the twenty-ninth degree of Capricorn. Jupiter completes a revolution in 12 years and has a deviation in latitude of 5°. Its apogee occurs in Virgo and its exaltation in the fifteenth degree of Cancer. Its ascents and descents prove that its orbit is eccentric to the earth. Saturn completes a revolution in slightly less than 30 years (elsewhere he had given it as 28 years) and has a deviation in latitude of 3° or even of only 2°. Its apogee is in Scorpio and its exaltation in the twentieth degree of Libra. The risings of Saturn are like those of the two other superior planets in that the rays of the rising and setting sun do not obscure it beyond 12°. First and second stations of Jupiter and Saturn occur 120° away from the sun on either side, and evening risings occur 180° away. The powerful effect of the sun's rays is responsible for the anomalies in the orbits of all the aforementioned planets and for their stations, retrogradations, and progressions. The rays strike the planets with such force as to elevate or depress them, cause them to deviate in latitude, or to retrogress. Readers will note the close correspondence between the accounts and figures of Pliny's and Capella's sections on planetary motions.

The most impressive features of Capella's book on astronomy are its logical arrangement of topics and its commendable regard for proportion. Only when he departs from the handbook framework and, like Macrobius, introduces elaborate computations for measuring planetary orbits, merely to amaze his readers, does he appear to have been carried away by his enthusiasms. The inadequacies of his comprehension and exposition of mathematical subjects have had to be scanted in this summary. Anyone with a knowledge of the elements of positional astronomy, after a reading of Geminus' *Introduction to the Phaenomena*, would be in a better position to understand Capella's account than was Capella himself.

The great authorities on classical astronomy and world geography in Western Europe during the early Middle Ages were Pliny, Solinus, Chalcidius, Macrobius, and Martianus Capella. The last three writers also introduced many readers to Pythagorean arithmetic; but as soon as Boethius had translated Nicomachus' *Introduction to Arithmetic*, his book superseded all others for serious students in this field. Capella's encyclopedia quickly became recognized as one of the most authoritative textbooks in the liberal arts. Fulgentius quotes from it early in the sixth century; Geoffrey of Tours refers to the author as "our Martianus" and calls the book an education in the seven disciplines. There are three extant commentaries, by John Scot Erigena, Remigius of Auxerre, and Dunchad or, according to the latest view, Martin of Laon. Notker Labeo (d. 1022), a monk of St. Gall who supervised the work of a group of German translators, made an Old High German translation, the last to be produced in any language. The precocious Hugo Grotius published an edition of the *Marriage* at the age of seventeen; and at one time the preparation of the Delphin edition was assigned to Leibniz, who was interested in restoring Capella to a position of esteem.

Inasmuch as the last book of Capella's encyclopedia, on music, avoids the usual mathematical approach to harmonic theory, it may be appropriately omitted from the present study.

ROMAN SCIENCE

IN THE

MIDDLE AGES

XII

CLASSICAL LEARNING UNDER THE OSTROGOTHS

For one brief spell, at the close of the fifth century and during the first quarter of the sixth century, Italy experienced a remarkable Hellenic revival, brought about by a combination of favorable circumstances and extraordinary men. Theodoric, son of one of the rulers of the Ostrogoths who were settled at that time in Pannonia, was sent as a boy of seven as a hostage to the court of Leo I at Constantinople to vouch to the eastern emperor for the good behavior of his father's subjects. The lad spent ten years at the capital city and absorbed indelible impressions of the advantages enjoyed by a well-ordered and richly cultivated Byzantine society. These impressions were to mold his own policies when, after his military victories over Odovacar in 493, Theodoric found himself ruler of Italy. His campaign against Odovacar had received the approval of the eastern emperor, Zeno, and he regarded himself as ruling by the sanction of Zeno. In his secure position Theodoric could have challenged Zeno by invading provinces under his suzerainty. Instead he settled down to provide his Gothic and Italian subjects with a regime blessed by amicable and equitable relations between conquering and conquered populations and graced by manifestations of the restoration of the Roman imperial polity and culture.

Theodoric inaugurated a vigorous program of improvements and reform, effecting large-scale public works, rehabilitation of the agricultural economy, and reduction of oppressive taxes. Roman imperial law was again in force, and the arts and letters were given substantial support. To Italians it must have appeared that the Golden Age had returned. By contrast with the havoc of the barbarian invasions of the preceding century, Theodoric's reign ushered in a period of peace and harmony seldom experienced before in Italy. For his countless benefactions to his subjects Theodoric won fame in legend and was commemorated in the Nibelungenlied as Dietrich von Bern. Among modern historians he has been regarded as the greatest of the barbarian conquerors and the spiritual forerunner of Charlemagne.

Outwardly Theodoric professed deference to the eastern emperor. Each ruler, eastern and western, designated a consul annually, and it appeared that the earlier order of a unified eastern and western empire had been re-established. Theodoric was intent however upon founding in Italy a strong national monarchy embodying the ideals, inculcated in his boyish mind at Constantinople, of law and order and a high level of culture. He envisioned an ideal society in which his Gothic settlers would furnish the strong arm of protection and his Italian subjects would provide experience in administration and the splendors of an ancient and glorious civilization. *Civilitas* became a favorite word with him, denoting the civil harmony resulting from the application of Roman law to Goth and Italian alike and a condition of religious toleration under which Arian invader and orthodox Catholic Italian and even Jewish sectarian enjoyed peaceful coexistence. Underneath, Theodoric was making a warrior caste of the Goths, and mercantile and laboring castes of the Italians; but at the top he showed a sensible deference of a barbarian toward a people gifted in matters of administration and jurisprudence, by assigning civil service positions, from the highest levels down, to Italian nobility and men of promise. He flattered the Roman Senate and made a pretense of consulting it on matters of highest importance.

Also implicit in Theodoric's conception of civilitas was a polite society enjoying the refinements of civilization. Although Theodoric, according to popular reports, was unable to sign his own name and used to indicate approval of documents by tracing upon them, with a metal stencil cut for this purpose, the word *legi*, "I have read this," he had a high regard for men of learning. To offset the handicaps of his meager education, he depended heavily upon the services of his secretary, whom he selected from the nobility for his outstanding intellect and learning. The two leading intellectuals of his kingdom, Boethius and Cassiodorus, became highly favored ministers, both attaining the consulship and the latter serving as his secretary. Both fervently extolled the virtues of Theodoric on public occasions and both enjoyed his encouragement in their efforts to revive classical studies after the barbarian incursions. During the short period when friendly relations obtained between the courts of Constantinople and Ravenna, there were many opportunities for cultural exchange, and a highly significant Greek revival was taking place in Italy. But near the close of Theodoric's reign, after his rupture with Justinian, caused by the latter's persecution of Arians in the east, Italy was again cut off from Hellenic contacts.

Boethius was slightly older than Cassiodorus, having been born about 480, into the illustrious and wealthy family of the Anicii. His father, Flavius Anicius Manlius Boethius, assumed the consulship in 487 but died soon thereafter; and the young Boethius was brought up and educated in some of the best homes of Rome. His guardian seems to have been Quintus Aurelius Memmius Symmachus, member of one of the most distinguished families of the later Roman Empire. Symmachus had served as consul in 485, and to him Boethius dedicated his first book. Boethius was happily wedded to Symmachus' daughter Rusticiana, a lady whose nobility of character and mind shone forth in the tribulations of her later life. Boethius was a man of brilliant intellect and exemplary character, born into favored circumstances, destined to distinguish himself in a political and literary career under a benevolent ruler—only to be cut down in his prime, at the age of forty-four, because he had fallen into the disfavor of Theodoric during the embittered years of that ruler's decline. Cassiodorus, who observed that Boethius equaled or surpassed the greatest writers of antiquity, was addicted, as we shall see, to fulsome praise; but his statement, as applied to Latin writers, is not extravagant. Gibbon's memorable characterization of Boethius as "the last of the Romans whom Cato or Tully [Cicero] could have acknowledged for their countryman"[1] is also a fitting one. As a writer on theoretical subjects Boethius deserves to be ranked with Varro, and as a philosopher with Cicero.

Had he not been born in a century when Roman learning was in an advanced state of decline, Boethius would have been a still more significant figure in the intellectual history of the West. With a keen perception of the dismal condition of contemporary science and philosophy, he saw the desperate need of basic texts. Much of the energy and enthusiasm of this talented man was expended upon the compilation of elementary textbooks and the translation of logical treatises. Early in life Boethius embarked upon a program, too ambitious for possible attainment, of providing scholars of his day with textbooks of all the mathematical disciplines and with translations and commentaries of all the works of Aristotle and Plato. It was his intention, at the conclusion of the latter project, to demonstrate that the philosophies of Plato and Aristotle were essentially in harmony with each other.

Boethius' intellectual gifts and aristocratic connections soon brought him to the attention of King Theodoric. Two letters, addressed to Boethius in 507 by Cassiodorus in the name of Theodoric, express the king's deep respect for him as a man of learning and scientific knowledge and charge him to carry out some earnest requests that call for the services of a man of his skill and attainments. The first of Cassiodorus' letters (*Variae* I.45), which is striking in its supposed significance, has been treated by nearly all specialists on Boethius as a crucial document in the reconstruction of his literary career. It is written in Cassiodorus' usual flattering style, but it is particularly flowery and ambiguous because Cassiodorus is trying to represent a youth of twenty-seven as a prodigious intellect who has placed all cultivated Italians under a debt of gratitude by making available to them the science and philosophy of the Greeks. When we come to observe Cassiodorus' own feeble efforts to embrace Greek mathematical subjects, it will be easy to understand why he was so appreciative of Boethius' talents.

The letter must be handled with extreme caution. One remark, "From afar you have entered the schools of Athens" (*Atheniensium scholas longe positus introisti*)—obviously figurative phraseology—was the basis of a legend which persisted for over a thousand years, that Boethius spent eighteen years studying in Greece. Gibbon, following a faulty textual reading of the phrase, was inclined to credit the tradition, though he regarded the term of eighteen years as too long.[2] The report is rejected on all sides now, and it is assumed that Boethius did not study in Greece. The letter goes on, in ornate and adulatory style, to point out that, thanks to Boethius' translations, "Pythagoras the musician and Ptolemy the astronomer are read as if they were Italians; the arithmetician Nicomachus and the geometrician Euclid are now available in Latin; the theologian Plato and the logician Aristotle discourse in the language of the Romans; [Boethius] has even restored to the Sicilians their mechanician Archimedes in Latin dress." Two, possibly three, of the items in Cassiodorus' statement had a basis in fact at the time of writing. Boethius had undoubtedly communicated to Cassiodorus his grandiose plans for the future. To suppose from this letter that Boethius translated Ptolemy's *Almagest* and Archimedes' works on mechanics is extremely rash.

The second letter, also of the year 507, though it too expresses the king's high regard for Boethius does not concern us directly here; but a third letter (*Variae* I.10) from Cassiodorus to Boethius in the name of Theodoric, written not much later, reveals the important responsibilities that Boethius was assuming early in his career. It requests Boethius, who was at the time count of the sacred largesses, to make amends to soldiers of the Praetorian Guard who had been paid in debased currency. In 510 Boethius was appointed to the consulship without a colleague; and in 522, at the height of his prosperity, he was promoted to master of the offices, in charge of the civil service system, with manifold responsibilities. The same year he escorted his two young sons to the Senate house to witness their investiture as joint consuls, and honored the occasion himself with a panegyric of Theodoric amid resounding applause.

Soon thereafter Boethius fell from Theodoric's favor because of his courage and resolute adherence to principles—at least, this is the traditional view about the circumstances of his downfall. When his friend Albinus, a man of consular rank, incurred the wrath of the praetorian prefect by trying to block a forced sale of grain that would have ruined the Campanian farmers, Boethius came to his rescue and thwarted the prefect. The immediate cause of Theodoric's anger against Boethius was the eastern emperor Justinian's persecution of the Arian heresy. Theodoric made an unsuccessful attempt to dissuade Justinian from his course and then engaged in measures of reprisal. It was believed that Justinian was trying to get the orthodox Italian population to throw off its Gothic Arian overlords. Letters allegedly written by Albinus, addressed to Justinian and containing treasonable proposals, were intercepted, and Albinus was charged with treason. The letters may have been forgeries. Other letters to Justinian, attributed to Boethius, are considered to have been forgeries. Boethius undertook Albinus' defense and, when it became evident that the king was trying to ruin the Senate by involving its entire membership in implications of treason, he courageously defended the senators. Both Albinus and Boethius were thrown into prison and suffered cruel deaths. So goes the traditional explanation of Boethius' fall from favor and his untimely death; but there are vexing questions, and it appears to some scholars that he may not have been free from guilt. Recently cogent arguments have been adduced to indicate that he was deeply implicated in religious and political controversies before his imprisonment.[3]

It was in prison, the feverish activity of his public life and intellectual pursuits behind him, that Boethius settled down to compose, in the solitude of his mind, his masterwork, *On the Consolation of Philosophy*. This deeply spiritual book, together with his partially completed collection of texts, was to make him one of the most influential writers of the early Middle Ages. But all of Plato's works and most of Aristotle's remained to be dealt with, and Boethius' great project was left unfulfilled. We can only imagine what a vast difference there would have been in medieval intellectual developments if Boethius had lived to an advanced old age and had been able to work steadily at his project of translating and commenting upon all the works of Plato and Aristotle.

Boethius' ambitious program of study and writing is unfolded in a letter in which he dedicates his treatise *On Arithmetic* to his father-in-law, Symmachus. This letter is appended to the treatise and serves as a preface. Boethius was twenty years of age when he wrote the work, and he refers to it in the letter as "the first fruits of my labor." He explains that his plan of study and writing was undertaken at the instance of Symmachus and that he is making available to Latin readers the riches of Greek literature. He frankly admits that *On Arithmetic* is a free version of Nicomachus' *Introduction to Arithmetic*, which he adds at the same time reduces the prolixity and clears up the difficulties of the original work. He feels that it is appropriate to begin with arithmetic because it is the first of the four mathematical disciplines. Martianus Capella had given the first position to geometry, following the example of Varro. Boethius adopts the attitude of Nicomachus, who, in the opening chapters of his *Introduction*, points out that arithmetic is the basis of the other mathematical sciences. In fact, it becomes evident from remarks in the letter that Nicomachus' opening chapters served as a guide to Boethius in his projected career. Nicomachus expresses the traditional Pythagorean view, first found in the *Republic*, that the only path to wisdom and philosophical truth is through the mastery of the four mathematical sciences, which he refers to as the four methods or ways. Boethius affirms that the heights of philosophy can be attained only by following "a sort of fourfold path" (*quodam quasi quadruvio*). This is the earliest use of the word *quadrivium* with reference to a course of studies. Boethius then announces his intention of compiling works on all four branches of mathematics that prepare one for the study of philosophy. It appears that he had completed his mathematical textbooks before he embarked upon his program of translating and commenting upon all the works of Plato and Aristotle.

Apuleius had translated Nicomachus' *Introduction* about the middle of the second century, but Boethius' work is regarded as independent of Apuleius. Comparison with the original reveals that he adheres to an intention, expressed in his preface, of not translating slavishly but condensing some portions of the original and introducing some minor additions. Nicomachus was his only significant source, and the additions are inconsequential: the insertion of numerical examples, for which there was little need; and the expansion of Nicomachus' statements of fundamentals and principles, which, though concise, are in themselves quite clear. It becomes evident that Boethius was not an expert mathematician and that, like other Latin translators and commentators, he was trying to exhibit his own talents in his departures from the original work. On the other hand, the severe criticism leveled at him by Moritz Cantor,[4] an eminent historian of mathematics, that he has omitted some of the essential features of Nicomachus' handbook, is without justification. Careful comparison of both texts led F. E. Robbins to the conclusion that Boethius' omissions do not detract from the value of his work as a translation.[5] Boethius' book held sway in the field of arithmetic in Western Europe, where it was known as "Boethian arithmetic," for over a thousand years. It also served as a basis for numerous Latin treatises on the subject.

Of the three great schoolmasters of the liberal arts in the early Middle Ages—Capella, Boethius, and Cassiodorus—Boethius was probably the most influential. Certainly in the field of music he was the dominant figure. His manual *On Music* was even more enduring than his treatise *On Arithmetic*. It continued to be used as a textbook in the universities in modern times, in Oxford as late as the eighteenth century. Capella had departed from age-old Pythagorean norms for the compilation of musical treatises by omitting a treatment of the numerical ratios underlying harmony. Boethius restored music to the mathematical quadrivium; he was largely responsible for the attitude of subsequent ages that the study of music was a proper one for theorists and mathematicians and was not related solely to the skill of instrumentalists, and that an analysis of numerical ratios was more reliable than mere appreciation of music dependent upon the fallible sense of hearing.

The five books of *On Music* have survived almost intact, only the last eleven chapters of Book V being missing. Boethius' main sources were Nicomachus' *Manual on Harmony* and Ptolemy's *Harmonics*, both extant, and Nicomachus' lost work *On Music*. Boethius' account of Pythagoras' discovery of the musical concords and of his theory of the harmony of the spheres follows Nicomachus' text very closely, but there are few other correspondences of extended length. Book I opens with a discussion of the effects of music upon mankind and classifies music as *mundane*, originating in the motion of the celestial sphere; *human*, combining incorporeal elements of the soul with corporeal elements of the body in numerical harmony; and *instrumental*, or music in its sensibly perceptible form. Boethius divides students of music into three groups, instrumentalists, composers, and theoreticians. Only the third group deserves to be considered musicians. Instrumentalists are mere performers—an ancient Greek attitude that superior accomplishments on instruments befitted slaves—and composers produce their works as a result of natural instincts and not knowledge. Book I also deals with the numerical ratios of consonance and dissonance, and points out analogies between musical harmony and planetary motions. Book II discourses upon Pythagorean number theory. Book III presents a criticism of the views of Aristoxenus, who had a practical approach to music. Book IV deals with the use of the monochord in the theory of music, and contains Euclidean echoes. Book V deals mainly with the musical theory presented in Ptolemy's *Harmonics*. Where Ptolemy and Nicomachus are found to be in disagreement, Boethius generally sides with Ptolemy. But in Book IV Boethius had so ineptly confused Ptolemy's use of terms in scales and modes that one critic doubted he knew Ptolemy's work at first hand.[6]

The third of Boethius' works on the mathematical quadrivium was a manual of geometry, of which only fragments remain. Cassiodorus flatly states, in his encyclopedia *On Sacred and Profane Literature*, that Boethius translated Euclid into Latin. A Latin work on geometry in two books, bearing Boethius' name and edited by Friedlein in 1867, was regarded by that editor as spurious and is so considered by historians of mathematics. Along with Euclidean material, it contains much practical matter on surveying and includes a discussion of the nine Hindu or Arabic numerals. This unauthentic work is believed to have been put together in the eleventh century.[7] However, it has some literal translations of Euclid's proofs and definitions that are faithful enough to have attracted the interest of textual critics. It has been surmised that these passages were extracted by pseudo-Boethius from the translation of Euclid prepared by Boethius. There is also a collection of manuscripts, dating from the tenth to the fourteenth centuries, bearing the title *Ars geometriae et arithmeticae*, that contains other fragments of a Latin version of Euclid thought to have been the one prepared by Boethius.

It is usually assumed that Boethius completed his mathematical quadrivium with a treatise on astronomy, but the evidence for such a work having existed is very meager. There is first the intention, expressed by Boethius in the Preface of *On Arithmetic*, to deal with all the mathematical disciplines in books of his own. Then there is that flattering and ambiguous letter of Cassiodorus, addressed to Boethius when he was about twenty-seven years of age, saying that, thanks to him, the musician Pythagoras and the astronomer Ptolemy have become available to Italians. Lastly, and most important, there is a brief letter written by Gerbert in 983 stating that he has discovered at Mantua a work on astrology in eight books by Boethius.[8] The work itself has left no known traces.

In completing the mathematical handbooks before he undertook his projected program of philosophical studies, Boethius was steadfastly adhering to Nicomachus' dictum that philosophical wisdom can be attained only after a mastery of mathematics. Despite the distractions of a busy career in politics and government service, he proceeded to produce a sizable collection of works on logic. He prepared a commentary on the *Categories* of Aristotle; a translation of Aristotle's *On Interpretation* and two separate commentaries on that work; translations of the *Prior* and *Posterior Analytics, Topics,* and *Sophistici Elenchi* of Aristotle; two separate commentaries on Porphyry's *Introduction to the Categories of Aristotle*; and a commentary on Cicero's *Topica*. He also published some logical works of his own: a treatise *On Categorical Syllogism* in two books and an *Introduction to Categorical Syllogism*, the main sources of which were Aristotle, Theophrastus, and Porphyry; *On Hypothetical Syllogism*, in two books; *On Division*, mainly derived from Porphyry; and *On Topical Differences*, in four books, drawn mainly from Themistius and Cicero. All of his logical works are extant, but not all of his translations.

His total collection of writings, comprised of mathematical, logical, and theological works, was destined to endear Boethius to religiously minded readers of the Middle Ages. Devout Churchmen would naturally have avoided most of the philosophical writings of the pagan classical authors, not wishing to be confused by them; but abstract mathematical studies and cold logic were safe subjects. Ever since the time of Clement and Origen, clergymen had been disposed to recommend mathematics and logic for their ancient function and role as propaedeutic subjects, to sharpen the wits of young men and to keep them out of mischief; but by now the pitfall and adversary was Satan.

To Gibbon, Boethius was the last of the true Romans. He was also the last notable Roman intellectual to reveal an intimate knowledge of the Greek language and learning. It is difficult in some fields of knowledge to draw a line between the end of the classical period and the beginning of the Dark Ages, but such is not the case in tracing the course of Roman science. Boethius and Cassiodorus, the two leading intellectuals of the West in the sixth century, were closely associated in spiritual and scholarly interests and their political careers were almost alike, except in ultimate outcome; yet in their comprehension of scientific subjects, as manifested in their writings, a wide chasm separates them.[9] After Boethius Latin writers on science had both feet firmly planted in the Middle Ages.

Cassiodorus, like Boethius, was born into a distinguished Roman family and had a father who attained to high political office. The date of Cassiodorus' birth has been placed at some time between 480 and 490, an earlier date within that period usually being favored. His career began as a legal assistant to his father, who, in 503, was elevated to praetorian prefect, the highest administrative post in the kingdom. Father and son served together until 507. About that time the younger Cassiodorus was invited to deliver a panegyric on Theodoric which so gladdened the king's heart that he appointed the young man quaestor of the Sacred Palace. The dignity of this position is indicated by its carrying with it the title of highest rank (*illustris*), accorded to only a few select ministers, and membership in the Senate. The quaestorship gave Cassiodorus an opportunity to demonstrate his talents as legal adviser, personal secretary, and diplomatic minister of the king. In drafting correspondence and official documents for Theodoric, he used a florid style that pleased the king, and he continued to be an enthusiastic supporter of the new order in which members of the Roman nobility accepted responsible positions in the benevolent regime of their foreign master. Cassiodorus' quaestorship appears to have ended in 512, and in 514 he was appointed consul ordinarius, by this time a titular office of high honor. He was subsequently enrolled in the Patriciate and served as governor of Lucania and his native Bruttii, perhaps during the period following his consulship. His extravagant praise of the Gothic nation and its line of kings in his *Chronicon* and *Getic History* had much to do with his being elevated to master of the offices in 523, a position which he retained until 527, a year after Theodoric's death. In 533 he was named praetorian prefect by the regent Amalasuntha, Theodoric's daughter, and held this responsible office with tactful agility and fidelity during the troubled years of the quick successions of Amalasuntha, Theodahad, and Witigis. In 537 Belisarius arranged to have his own appointee inducted as prefect; it was probably about this time that Cassiodorus was seriously considering retiring from public life and devoting his remaining years to monastic service and scholarship.

After Justinian's expeditionary force, under the vigorous command of Belisarius, had quickly subjugated the Vandal kingdom in North Africa, the eastern emperor found the pretext he had been seeking to justify moving against the Ostrogoth regime in Italy. Amalasuntha had married Theodahad, who shortly thereafter had her imprisoned. When she appealed to Justinian for aid, Theodahad put her to death. Belisarius was ordered to invade Italy, using North Africa as his base of operations. His campaign ended successfully in the capture of Ravenna in 540. With the crushing of Ostrogothic rule in Italy, Cassiodorus' political and administrative career would surely have been terminated. His life, almost since boyhood, had been one of dedicated service to a regime that seemed to him a blending of the best traits of Goth and Roman. With its destruction there was nothing left for him to do but to find satisfaction within himself.

He returned to the beloved haunts of his childhood and founded a double monastery near Scyllacium, his birthplace in the south Italian province of Bruttii, organizing the settlement according to a rule that was more strongly influenced by Cassian than by Benedict. The surroundings of the monastery must have been attractive. The settlement consisted of a hermitage in a valley of Mount Castellum, and a less austere retreat with watered gardens and an artificial fishpond from which the monastery took the name Vivarium. Bathing facilities were constructed at a nearby stream for the therapy of ailing brethren. The setting is fondly described by Cassiodorus near the close of the first book of his encyclopedia *On Sacred and Profane Literature*.

The decision to retire to a monastic life was not a sudden one. During the years when affairs of state were in an upset condition because of court intrigue and weakling control, Cassiodorus was becoming increasingly concerned with measures to combat the grievous conditions of religious scholarship in Italy. He had been distressed by the destruction of precious manuscripts in the barbarian invasions and by the fact that, although it was still possible for the sons of wealthy men to get an adequate secular education from Greek tutors, there was no institution where a religious education could be provided for aspirants to the priesthood, as there was in other countries. In 535 or 536 Cassiodorus had consulted with Pope Agapetus about the feasibility of establishing a theological college at Rome. Then, with the fall of the Ostrogothic kingdom, he embarked upon a second career, which was to occupy him until his death at past ninety and was also to have profound influence upon intellectual life in the West.

At his monastery at Scyllacium, Cassiodorus drew together a company of monks whom he set to work copying and translating manuscripts. He himself was zealously accumulating in the monastery library manuscripts that he and others had been gathering from three continents. The basis of all education, he maintained, was the Bible; and he exhorted his monks to seek enlightenment in study of the scriptures and commentaries upon them, first Latin commentaries and, if need be, Greek and Hebrew as well. He engaged his scholar-monks in a project of emending and collating biblical texts and translating commentaries and ecclesiastical histories. At one point in *On Sacred and Profane Literature* he notes with pride that a Vivarian version of Josephus' learned *Jewish Antiquities* has been turned out in twenty-two books, with great labor. It was while he himself was occupied with compiling a huge commentary on the *Psalms*—over a thousand columns in the Migne edition—that he decided to prepare a manual of study for his monks.

This manual, the full title of which is *On Training in Sacred and Profane Literature*, was not considered by Cassiodorus one of his important publications, being a rather scanty book of instructions and advice designed for scholars at Vivarium. For us, however, in reconstructing his scholarly milieu and motivations, it is his key work. It also served in subsequent ages to establish Cassiodorus as one of the greatest of the postclassical encyclopedists and one of the most influential intellectual figures of the Middle Ages. Considering his attitude that the Bible was the foundation of all knowledge and that training in secular subjects was but a means to the end of understanding the scriptures better, we would expect the subject matter of Book I, sacred literature, to hold greater interest for Cassiodorus than that of Book II, profane literature. Book I outlines a program of scholarly activity for the monks, who are fervently exhorted to diligent study and devout meditation. Cassiodorus gives evidence of his own enthusiasm for religious scholarship in his pride in the library manuscript collection and his assessment of the merits of various commentators on books of the Bible. He also reveals an open-mindedness about scholarship. He points out that even heretical writers may be studied with profit and that biblical students will find a knowledge of such subjects as geography and orthography helpful in interpreting scriptures.

Book I contains thirty-three chapters, a number corresponding to the age of the Master when he was crucified. Book II, with its seven chapters, is declared to be an addendum to Book I. Each of the chapters is devoted to one of the traditional seven liberal arts. There is scriptural significance in the very number of the seven secular arts. Did not David say that he praised the Lord seven times a day; did not Solomon speak of a house of wisdom built of seven pillars; did not the Book of Revelation refer to the seven churches, the seven Spirits, the seven golden candlesticks, and the seven stars? Arithmetic is a commendable study for clerics. Did not the Lord say, "But the very hairs of your head are all numbered"?

Cassiodorus' plea on behalf of liberal studies had lasting effects, some of which would have surprised him. It has been suggested that he was induced to compile this secular manual because of his displeasure with Capella's flamboyant handling of the liberal arts in his encyclopedia. He refers twice to Capella's work, the second time to say that he has heard of its existence but has been unable to obtain a copy. Because of its unavailability, he feels called upon to offer his own discussion, poor as it may be, of the mathematical quadrivium. Instead of supplanting Capella's book, Cassiodorus' handling of the liberal arts gave wider acceptance to these studies in religious institutions and added greatly to the currency and popularity of Capella's work. Although, in Cassiodorus' eyes, Book II was of minor significance in comparison with Book I, it was the secular portion of his work that had greater influence upon later intellectual developments. Bede, Isidore, and Alcuin were familiar with Book II but did not know of the existence of Book I.

Book II of Cassiodorus' encyclopedia is on the whole a very concise treatment of the seven liberal arts. The attention devoted to the various subjects is disproportionate. His favorite subjects, rhetoric and dialectic, occupy slightly more than half the book; and grammar, geometry, and astronomy get very meager treatment. The body of each chapter contains stock handbook material, extracted or digested from earlier writers, to which Cassiodorus adds a preface or conclusion containing observations of his own, mostly to show a connection between the subject at hand and some aspect of the spiritual life.

Seldom is a Latin compiler in the Middle Ages found to be as revealing and honorable about his sources as Cassiodorus. He cites authorities frequently, on an average of more than one to a page; and most of the time his citations are reliable. Occasionally he inserts the names of important authorities whose works he evidently did not handle or possess in his library, but his intention is not, as in the case of most Latin compilers, to impress or deceive his readers. He is merely offering bibliographical aids to students who may become absorbed in a subject and perhaps encounter the books elsewhere. Occasionally he lapses into the careless practices of compilers: he cites the same work by different titles; sometimes he cites an original work when he has been using a Latin translation; several times he reproduces or paraphrases without acknowledgment; and on one occasion he cites Cicero's *On the Orator* for a statement which actually came from the anonymous *Rhetorica ad Heren-*

nium. Donatus furnished the definitions for Cassiodorus' short chapter on grammar. The chapter on rhetoric was drawn chiefly from Cicero's *On Invention* and also from Fortunatianus, whose work on the subject is extant. The chapter on dialectic was derived from Aristotle's *Categories*, pseudo-Apuleius' *On Interpretation*, three works by Marius Victorinus, and from Boethius' translation of Porphyry's *Introduction* and his commentary on Aristotle's *On Interpretation.* Cassiodorus also made free use of quotations from Cicero's speeches to serve as illustrations of his definitions.

His extensive use of Cicero's works in his rhetorical trivium would lead us to suppose that Cassiodorus had gotten off the beaten path of the hackneyed manuals on the liberal arts and was not producing another mechanical digest or compilation from earlier manuals, although he naturally had to depend on standard handbook sources for most of his material. This impression is confirmed by what little we know about the sources of his mathematical quadrivium. Here, too, he made some use of works that were not the common property of compilers. He refers to Varro's encyclopedia six times, but nowhere does he say that he possesses a copy nor does he recommend it to his monks, and it is likely that he derived his quotations secondarily. Capella he seems to have known only by reputation. Inasmuch as mathematical subjects were not nearly so popular as rhetoric and dialectic in the Middle Ages, fewer ancient mathematical works have survived, and consequently it is much more difficult to trace the sources of Cassiodorus' chapters on the quadrivium.

In his chapter on arithmetic Cassiodorus recommends for further reading Nicomachus' *Introduction* and the Latin translations of it by Apuleius and Boethius. All of the Nicomachean material he handles is to be found in Boethius' translation; consequently, there is no reason to suppose he consulted the original work in the preparation of this chapter. He presents in abstract three of the four major divisions of Nicomachean arithmetic: the classification of numbers and their division into odds and evens, abundant, deficient, and perfect; the relations of numbers to each other; and figured numbers: linear, plane, solid, cubical, etc. The Nicomachean discussion of proportions is not included. Cassiodorus concludes the chapter appropriately by suppressing the conventional arithmological discussion of the mystical powers and associations of the numbers of the Pythagorean decad, substituting for them religious associations. The number one refers to the single God and one

Lord, two to the Testaments, three to the Trinity, four to the Gospels, five to the Pentateuch, six to the day on which the Lord made man in his own image, and seven to the sevenfold Holy Ghost. He passes on to the subject of music without completing the decad.

From the disproportionate amount of attention that Cassiodorus devotes to arithmetic and music, it is obvious that these were his favorite mathematical subjects. In the paragraphs of his own composition in these chapters he shows some enthusiasm in relating the subjects to the divine order; but the bodies of these chapters, containing mostly conventional handbook material, turn out to be meager classifications, divisions, and definitions of terms.

The chapter on music opens with some trite remarks, reduced to pithy sentences, about Pythagoras' experiment with the hammers and the Muses' connection with the beginnings of music. Then follows a paragraph pointing out the importance of music to a religious man. The science of music deals with harmonious relationships: the righteous man is attuned to these harmonies; a life of sin is one dissociated from music. Another example of the relation between music and religion is found in the ten-stringed instrument of the Decalogue. The musical divisions and definitions follow. Music has three divisions: harmonics, rhythmics, and metrics. There are three types of musical instruments: percussion, stringed, and wind. There are six consonant intervals: the diatessaron, or fourth; the diapente, or fifth; the diapason, or octave; the diapason and diatessaron, or octave and fourth; the diapason and diapente, or octave and fifth; and the disdiapason, or double octave. And lastly, there are fifteen different kinds of tones. After outlining the traditional handbook materials on music, Cassiodorus resorts to his customary practice of concluding the chapter with an expression of further personal opinions. Stories of the musical powers of Orpheus and the Sirens are rejected as fables; but readers are reminded of David's success in cleansing Saul's spirit by musical ministrations, and of the report of Asclepiades' restoring a madman to sanity by means of music.

In the bibliography for his chapter on music Cassiodorus refers his monks to the Greek works of Alypius, [pseudo-] Euclid, Ptolemy, and Gaudentius (this last also cited in the Latin translation of Mutianus), and the Latin work of Albinus, which he recalls having read at Rome and hopes has not been destroyed by the latest barbarian invasion. He also lists Latin works by Apuleius, which he says he does not possess;

Augustine's *On Music*, in six books; and Censorinus' *On Accents*, which is no longer extant. Pierre Courcelle, who has made a careful investigation of Cassiodorus' possible sources for this chapter in *Les lettres grecques en Occident de Macrobe à Cassiodore*, declares that he made no use of four extant works listed in his bibliography: Boethius' treatise *On Music*, pseudo-Euclid's *Introduction to Harmony*, Ptolemy's *Harmonics*, and Augustine's *On Music*. On the other hand he did avail himself of the lost Latin translation of Gaudentius, as resemblances with Gaudentius' extant work indicate. Cassiodorus' six *symphoniae* or consonances could have been derived from Gaudentius, but his fifteen tones are not found in Gaudentius' work. By a remarkable coincidence—one of many that Courcelle has discovered in his excellent book—the one fragment of Alypius' *Introduction* to survive contains those fifteen distinctions of tones. Since the conventional division is into thirteen tones, and there are other correspondences between Alypius and Cassiodorus present in that fragment, Courcelle concludes that Cassiodorus introduced into his chapter excerpts from Alypius' lost work.[10]

Cassiodorus shows almost no interest in geometry and devotes little more than a page to it. He opens with brief remarks about the origins of geometry, noting that some authorities assign its beginnings to the Egyptians. Then he cites Varro on the early use of geometry in determining property lines, and its later uses in measuring the earth's dimensions and the distances of the moon, sun, and sky. He recommends Censorinus' book *On His Birthday* for an account of the dimensions of the earth and universe. Courcelle believes that Cassiodorus derived his Varronian statements from Censorinus. Then follows a fourfold division of Euclidean geometry into plane geometry, numerical magnitude, commensurables and incommensurables, and solid geometry. He closes by referring to three Greek authorities on geometry—Euclid, Apollonius, and Archimedes—and by praising Boethius' translation of Euclid's *Elements*.

At the opening of his chapter on astronomy Cassiodorus recommends Seneca's *On the Shape of the World*, a copy of which he says is in the monastery library. His recommendation here is strong evidence of the high regard in which that pagan philosopher was held by Christians—for a little later he warns his readers to eschew human opinions on this very subject and follow the scriptural account. It becomes clear that the core of Cassiodorus' astronomical matter was not Senecan but Ptolemaic in origin. He opens with personal remarks about the wonders of the sky and the joys and rewards of contemplating the heavens. The movements of celestial bodies were established by the Creator. There have been miraculous phenomena, as when Joshua requested that the sun stand still, when the nativity star appeared to the wise men, and when the earth was darkened for three hours at the Crucifixion. Then he inexpertly reduces the subject of astronomy to sixteen divisions—including East, West, North, and South (!)—and does not even observe a logical order in listing them. His technical terminology is Ptolemaic, and widely dispersed echoes of Ptolemy's bulky *Almagest* can be pointed out; but it is much more reasonable to assume that Cassiodorus' ultimate source was some reduction of Ptolemaic astronomy, such as Ptolemy's *Planetary Hypotheses* or his *Hand Tables* (he cites the latter), and that his immediate source was Boethius' lost manual on astronomy, which Boethius in turn had derived from some Ptolemaic manual. Cassiodorus then lists the seven geographical latitudes, from Meroë to the mouth of the Borysthenes, and closes with some remarks about writings of the Church Fathers on the subject of astronomy: Basil has dealt with the subject in his *Six Days of the Creation*, and Augustine warns against the perverted use of astronomy to predict futures; scriptural authority as to the shape of the world, and not Varro's conception, is to be relied upon.

Cassiodorus was the most influential of the Churchmen in persuading clerics of the advantages accruing from studying the pagan disciplines. He could have reasonably argued that grammar improved their written style, rhetoric and dialectic developed their powers to controvert the arguments of heretics and unbelievers, arithmetic and astronomy assisted them in computing movable feast days, and harmonic theory added to their appreciation of the musical forms of worship. There has been heated controversy as to whether or not the Benedictine Rule exerted an influence upon the organization of monastic life at Vivarium;[11] but it does appear likely that Cassiodorus had a powerful influence upon the Benedictine Order in helping to cultivate their intense interest in secular letters and learning. Of prime importance is the incalculable debt that the entire Western world of letters owes to him. If he had not instituted a practice among his monks of transcribing copies of manuscripts, the bulk of ancient Latin literature that has survived would have been lost to us.

The Greek revival in the West was a short-lived one. It began propitiously, with the earnest encouragement of Theodoric and with wide-open contacts between Italian capitals and Constantinople. The brilliant and eager young Boethius had avowed that he would translate and comment upon all the writings of Plato and Aristotle, but early death terminated his project. Cassiodorus, living into his nineties, had access to at least as many manuscripts of Greek classics as his short-lived colleague but he had only a small measure of Boethius' enthusiasm for academic subjects and no mathematical ability at all. Thus a movement which, in its vigorous early stage, produced a new translation of Nicomachus' *Introduction*, an abridged translation of Euclid's *Elements*, a translation of some Ptolemaic handbook on astronomy, and a durable manual on music which included an elaborate account of Ptolemy's *Harmonics*, faded almost to the vanishing point in a set of threadbare classifications and definitions compiled by an elderly man whose interest it was to prepare reading lists for his younger monks and to point out the connection between a knowledge of pagan disciplines and the spiritual life.

XIII

ENCYCLOPEDIC SCIENCE IN THE BORDERLANDS

Of the four great exponents of pagan encyclopedic learning in the early Middle Ages, Boethius and Cassiodorus were intimate colleagues in Italy, representing the Greek revival that occurred during Theodoric's reign; Isidore and Bede, on the other hand, were widely separated in time and place, the one flourishing in Seville in the first decades of the seventh century, the other in a northern district of England about a century later. At first glance it might appear that the culminations of early medieval learning which these men represented in Italy, Spain, and England were sporadic and discrete phenomena; but it is not hard to trace a connection.

Cassiodorus, Isidore, and Bede were all devout Churchmen; and during this period the Catholic Church, in its determined struggle to obliterate heresies and heathenism throughout Western Europe, maintained close contacts with the outposts of the faith. One such spiritual contact, between Gregory the Great, later to become Pope Gregory I, and Leander, older brother of Isidore, was to have important effects in advancing the career of Leander and the cause of Catholicism in Spain. Many years afterward, when Gregory was Pope and Leander was bishop of Seville, Gregory dedicated to Leander a commentary on the Book of

Job bearing the title *Morals*, a work in thirty-five books which was the fruit of many years of laborious study. Shortly thereafter, the same Gregory initiated one of the most significant enterprises ever undertaken by the Church: he dispatched two missionary parties, one under St. Augustine in 596, the other under Mellitus and Paulinus in 601, to convert the heathen of Britain. This missionary movement, indelibly influenced by its Benedictine origins and ties, was later to develop famous centers of religious education and traditional learning in England, the most remarkable scholarly product of which, in the early period, was the Venerable Bede. Cassiodorus, too, was a prominent factor in the mingling of religious and secular instruction in the Church schools of Western Europe because he, above all others, gave respectability to the classical liberal arts in the eyes of Churchmen.

The family into which Isidore was born was a prominent Hispano-Roman family of Cartagena. The names used by its members were Roman. It is thought that his father, Severianus, was an important official of Cartagena who, having gotten into difficulties, was forced to leave Cartagena and removed his family to Seville. Isidore was born about the time of the removal (c. 570). Severianus died shortly afterward, and the responsibility of rearing and educating the young lad Isidore fell upon his brother, Leander. Leander is said to have been stern and rigorous in his upbringing of Isidore. That he schooled him thoroughly and inculcated industrious habits in him there cannot be the slightest doubt. Leander was himself a scholar of high repute.

Shortly before Isidore's father moved to Seville, Leovigild came to the Visigothic throne in Spain. He and his son Reccared were the greatest of the Visigothic kings in Spain. King Leovigild began his reign (567) with a feverish display of energy, overcoming both rivals at home and enemies on the borders and throwing off all pretense of allegiance to the eastern emperor. He remained, however, a staunch adherent of the Arian faith. He gave important responsibilities to his sons Hermenegild and Reccared, making the former governor of Baetica and marrying him to a Frankish princess, Ingundis. She and Leander, who was by this time bishop of Seville, succeeded in converting Hermenegild to the Catholic faith; when Hermenegild revolted against his father, Leander prudently went into exile. Hermenegild was subdued and executed. Reccared succeeded his father Leovigild in 586, and three years later was converted to the Catholic faith at the Third Council of Toledo.

Gregory the Great was overcome with joy as he wrote to Leander, congratulating him upon winning Reccared and the Visigothic kingdom to the support of the Church. Death in 599 terminated the illustrious career of this vigorous champion of the faith who later became St. Leander. Shortly after his death his brother Isidore succeeded him as bishop of Seville.

If Isidore's birth is assumed to have been close to the year 570, he must have been admitted to serve as deacon and priest at near the minimum canonical age, and he was only about thirty when he succeeded his brother as bishop. It is surprising that so few details are known of the life of as prolific a writer and as energetic and eminent an officer of the Church as Isidore. He presided at a council in Seville in 619 and at the Fourth Council of Toledo in 633, and served as bishop until his death in 636. He dedicated an early work, *On the Nature of Things*, to the Visigothic king Sisebut (612–620); and maintained a firm friendship and scholarly correspondence with his former pupil, Braulio, bishop of Saragossa.

In his abounding energy, industrious habits, and breadth of learning, Isidore may be favorably compared with Pliny. It might appear from his theological writings and the ecclesiastical documents attributed to him that he spent all his wakeful hours during his tenure as bishop attending to matters of Church organization and administration and settling even minor routines and liturgical observances. Yet, judging from his secular writings, he seems to have devoted more time to these studies than to ecclesiastical matters. He founded a school at Seville and probably taught there. His publications, occupying three volumes of Migne's *Patrologia latina*, compare with Pliny's in bulk. Braulio lists seventeen separate works by Isidore, the last being his *Etymologies* or *Origins* in twenty books; and the list is incomplete. Considering the low ebb of scholarship in his day and the fact that leading members of the Church, like Pope Gregory, were outspoken opponents of secular learning, it would be unfair to focus attention upon Isidore's ineptitudes as a scholar and his unoriginality as a compiler without first acknowledging his importance in the history of scholarship. Viewed against the disheartening background of contemporary letters, Isidore's *Etymologies* is one of the outstanding feats of scholarship of all time. The collection of manuscripts available to him in the library at Seville must have been a sizable one, yet it did not include a copy of Pliny's popular and widely circulated

Natural History. The thirty-seven books of Pliny's work gave systematic treatment to a restricted number of subjects; Isidore's encyclopedia, in twenty books, begins with extended discussions of all seven of the liberal arts and encompasses within its universal scope even such minor topics as amusements, garments, and household utensils. Merely as a compilation the *Etymologies* is a momentous accomplishment. It was one of the most widely read books for the next thousand years; and its serving as a model for encyclopedists as late as Vincent of Beauvais affords a measure of Isidore's importance in the world of scholarship.

In another age and a milieu more favorable for research, Isidore's scholarly sins might have outweighed his virtues. He employed all the most reprehensible practices of Latin compilers to a marked degree. His scientific and scholarly experience was limited to books; no original contributions or firsthand observations were recorded in his works. His secular compilations were reiterations of classical learning, and he himself was far removed from an understanding of classical times and events. To Isidore the astronomer Ptolemy was king of Alexandria and Hannibal was the general who sacked Corinth. The Roman satirist Petronius, courtier of Nero and author of a ribald satirical novel, the *Satiricon*, in representing his character Trimalchio as a ridiculous parvenu had him explain to his banquet guests that Hannibal pillaged Corinth immediately after his capture of Troy. By Isidore's time this bit of buffoonery was taken for fact.

Meticulous research has uncovered abundant deceptions in Isidore's references to his authorities. Kettner has pointed out that 36 passages citing Varro were not derived from Varro and that Isidore did not consult Varro's works. Klussmann found nearly 70 passages derived from Tertullian's works, and yet Isidore does not cite Tertullian.[1] Mynors' edition of Cassiodorus' *On Sacred and Profane Literature* lists 34 borrowings from that slim work in the first three books of the *Etymologies*, many of them extended passages of close correspondence or verbatim copying; but Isidore withholds acknowledgment.[2] Mommsen, in his edition of Solinus' *Collectanea*, lists nearly 600 excerpts drawn by Isidore, but Solinus' name appears nowhere in the *Etymologies*. Festus, another writer who is presumed to have been one of his main sources, is not mentioned. On the other hand, there is an array of citations of early Latin playwrights, satirists, and epic poets, most of whose works had disappeared centuries before Isidore's time: Livius Andronicus, Naevius, Plautus, Ennius, Caecilius, Turpilius, Afranius, Pacuvius, and Lucilius. It has been presumed that Isidore derived most of these excerpts from Servius' Virgil commentaries, although there have been vigorous protests against this view. Servius appears to have been one of Isidore's chief sources, but his name is omitted.[3] It is not reassuring to reflect that our knowledge today about the texts and attributions of the collected fragments of lost classical works must depend, in many cases, upon the scholarship of such compilers as Isidore.

Only two of Isidore's works are of interest to students of the history of science: his bulky *Etymologies* and a slender earlier work entitled *On the Nature of Things.* The former purports to be an etymological dictionary. The less said about Isidore's word derivations, the better. One already noted—connecting the name of the Isle of Thanet, where snakes allegedly could not survive, with the Greek word for death, *thanatos*—is typical. The book should be judged as an encyclopedia and not on the basis of the etymologies. Isidore knew no Greek, and when he occasionally inserts Greek words or phrases they are seen to be commonplaces of the sort that had been transmitted by generations of Latin compilers who also knew little or no Greek. The first three books of the *Etymologies* are devoted to the seven liberal arts, and Books XIII and XIV deal with cosmography and physical geography. As might be expected, the rhetorical trivium receives much more attention than the mathematical quadrivium. All of Book I's 56 pages are on grammar. Rhetoric and dialectic share the pages of Book II equally. The subjects of the mathematical quadrivium are cramped into Book III, with 10, 4, 6, and 16 pages allotted to arithmetic, geometry, music, and astronomy respectively.

The order in which the medieval encyclopedists took up the subjects of the mathematical quadrivium was not a set one. Capella's order was geometry, arithmetic, astronomy, and music. Cassiodorus made no use of Capella's book; he depended for his mathematical rudiments mainly upon Boethius, and naturally subscribed to Boethius' attitude that arithmetic was the basic mathematical study and should be given first place. Because he had greater interest in music than in geometry and astronomy, he took up music second and left geometry and astronomy for third and fourth positions. Isidore's order was arithmetic, geometry, music, and astronomy. The relegation of astronomy to last position by the Church Fathers was largely because of its suspected association with astrology, casting of nativities, and predictions of the future.

The first nine chapters of Book III of the *Etymologies* contain Isidore's discussion of Nicomachean arithmetic. It is essentially a reduction of Cassiodorus' chapter on arithmetic, which, in turn, was presumably derived from Boethius' translation of Nicomachus' *Introduction*. Cassiodorus had displayed little interest in the subject and no competence as a mathematician. Isidore shows even less interest. His preface is copied from Cassiodorus, as are some long sentences in the body of his discussion. There are correspondences of shorter extent throughout. Isidore uses fewer examples than Cassiodorus and omits or curtails much of Cassiodorus' account; but there is also some Nicomachean material in Isidore's treatment that is not found in Cassiodorus': some examples of perfect numbers in Chapter 5 and a brief discussion of arithmetical, geometrical, and harmonic means in Chapter 8. The introduction of additional Nicomachean material should not lead us to suppose that Isidore used Boethius' or Apuleius' translation to supplement Cassiodorus' discussion. He did not have sufficient interest in the subject to consult a full-scale treatment in a handbook. Rather it is to be assumed that the Nicomachean examples and definitions were available in some handy abridged form.

During this age all that was left to indicate some scant interest persisting in the study of geometry was a few Euclidean classifications and definitions filtering down from the works of Capella and Boethius. Capella had planned to give each of the liberal arts elaborate treatment in a full-length book. To extend his book on geometry to normal size he was obliged to insert a lengthy account of world geography, extracted from the works of Pliny and Solinus; and his ten-page discussion of geometry thus became an addendum. Cassiodorus did not introduce a geographical account into his disciplines; consequently his chapter on geometry is not much more than a page in length. Isidore appropriated the geographical materials of Solinus, but in two separate books later in the *Etymologies*; his treatment of geometry is also very scanty.

He devotes five brief sections to the subject. The first two, on the origins and uses of geometry and its fourfold Euclidean divisions, substantially reproduce Cassiodorus' materials. The other sections, containing definitions of geometrical figures and terminology, are not found in

Cassiodorus, but were obviously derived from some handy summary and not from a Euclidean manual.

Astronomy was a ticklish subject for patristic writers to handle systematically. That the Lord God, and not some Platonic Fabricator, had created heaven and earth and ordained the movements of the stars and planets was absolutely clear; but when it came to the shape of the universe and earth, it was not always easy to reconcile scriptural concepts with the doctrines of pagan philosophers and astronomers. Very few Church Fathers who wrote on the subject were as bitter as Lactantius in his denunciation of such ridiculous pagan notions as a spherical earth and the existence of antipodeans. Their attitude was rather one of caution against close study of pagan doctrines, and their own writings on the physical universe were full of ambiguities. Any Father who gave only a cursory or gingerly perusal to a work on classical astronomy would find little difficulty in reconciling the figurative and allegorical descriptions of the scriptures with the abstruse cosmographies of pagan astronomers. To be sure, even classical Latin encyclopedists and handbook compilers who vaunted their knowledge of scientific subjects and seemingly applied themselves to such studies were seldom able to comprehend the elements of Greek astronomy and mathematical geography, and mingled bits of doctrines belonging to different periods. Should we be surprised if, in patristic accounts, we find the heavens shaped like a tabernacle, tent, vault, or sphere, and the earth represented as a plane, disc, cylinder, or sphere—a mélange of scriptural concepts and Platonic, Lucretian, and even pre-Socratic doctrines?

Let us consider the attitudes of patristic writers on such matters. By Isidore's time, Cassiodorus' cosmographical views had become representative of the opinions of the more enlightened clergy, and he enjoyed the respect and esteem of educators in the Church schools. He recommended Ptolemy as an authority on astronomy, undertook with assurance to define several of Ptolemy's technical terms; spoke briefly on the relative sizes of the sun, moon, and earth; listed the seven geographical latitudes and pointed out that sundials have to be corrected to function properly in each of these latitudes; and referred to the hemisphere above the earth and the hemisphere beneath the earth; yet nowhere did he say that the earth has a spherical shape. On this subject he considered it absurd to follow human opinions when divine pronouncements had been made, and he recommended St. Basil's *Six Days of the Creation* and St. Augustine's *On Christian Learning*. He did not present Basil's and Augustine's views, but we find them temperate, moderate—and equivocal. Basil understands that stars in the south polar region are invisible and that the sun's course through the northern and southern halves of the zodiac produces summer and winter, but he finds it hard to credit a spherical universe in the light of the Genesis concept of upper waters, and he cannot decide whether the earth is a sphere, disc, or cylinder. Moses had nothing to say about this or about an estimate of the earth's dimensions as 180,000 stades. Augustine too is troubled by the waters above the firmament, the question of the sphericity of the earth, and the existence of antipodeans. Still another favorite authority of Isidore on the subject of cosmography was St. Ambrose, Bishop of Milan. Ambrose assumes a spherical universe and, when he is confronted with the waters outside the sphere, he explains that a house can be round on the inside and square on the outside. Or perhaps the waters are suspended freely in space.

In the welter of such confused and equivocating authorities we should not be surprised if Isidore's conceptions are sometimes childishly naive and often inconsistent. Certainly consistency is not to be expected of a medieval cosmographer who was conflating scriptural and secular authorities, when the ancient Greek handbook astronomers are found to have had no compunctions about embracing irreconcilable doctrines. Isidore's cosmography is presented in the second half of Book III and in Chapters 4–6 of Book XIII of the *Etymologies*, as well as in *On the Nature of Things*, which is some eighty pages in length. His most important sources are thought to have been Suetonius' lost *Pratum*, the scholia to Germanicus' *Aratea*, Hyginus' *Astronomicon*, Ambrose' *Six Days*, the *Recognitiones* of pseudo-Clement, several works of Augustine, and, of course, the Bible.

Isidore's universe is spherical, it makes a daily revolution, and is the abode of the fixed stars. The earth is at the center of the universe. He speaks vaguely in the *Etymologies* about the seven heavens of the seven planets, ascribing the conception to "the philosophers"; but in *On the Nature of Things* he reveals that this statement of the philosophers is a quotation from Ambrose' *Six Days*, associating multiple heavens with a verse in the *Psalms* (148:4), and he goes on to say that even reckless men would not presume to know the number of the heavens. The planets' motions produce harmony and are in a direction contrary to that of other celestial bodies. Isidore states that the sun is larger than the earth and the earth larger than the moon, and gives the correct explanation of eclipses of the sun and moon, information which he could, and may, have gotten from Cassiodorus. He lists the same geographical latitudes as Cassiodorus. The sun completes a revolution in a year, being at apogee in its northern tracts and at perigee in its southern tracts. Conjunctions of the moon occur on the thirtieth day. The order of the planets is moon, Mercury, Venus, sun, Mars, Jupiter, and Saturn. Isidore's concise statement in the *Etymologies* about the planets' periods is misleading. He says that the moon completes its course in 8 years, Mercury in 20, Venus in 9, the sun in 19, Mars (which he calls "Vesper") in 15, Jupiter (which he calls "Phaethon") in 12, and Saturn in 30. In *On the Nature of Things* he more helpfully explains that these figures refer to the periods required for the planets to complete their courses and return to their former precise positions in the constellations. In any case, Isidore was merely copying statements from earlier authorities and did not comprehend what was involved. The planets are so called because they stray through the entire universe in their erratic motions. When they subtract degrees, they are said to be in retrograde motion; when they stand still, they are in station. He attributes the stations and retrogradations to the powerful effect of the sun's rays, an old Stoic notion, and quotes the Roman poet Lucan as an authority. There are also some discrepancies in stellar phenomena. Some stars set quickly and rise slowly, others rise quickly and set slowly. Much of the astronomical material in the *Etymologies* is copied from his earlier work *On the Nature of Things*, which in general offered a better and fuller discussion.

Against a background of the confusion and perplexities of the Church Fathers with regard to the shape of the earth, it is interesting to examine Isidore's views on the subject. In chapter 10 of *On the Nature of Things* he begins his discussion of mathematical geography in the manner of a Greek compiler, by describing the five celestial circles or parallels and using an excellent illustration of the five fingers of the hand, fully extended from the open palm, the thumb representing the arctic circle, the index finger the summer tropic, the middle finger the equator, the fourth finger the winter tropic, and the little finger the antarctic circle. By this time Isidore reveals that he has not understood what is involved in his borrowed illustration. It turns out that he is not really describing celestial circles but circular zones on the flat surface of the earth. The equatorial circle will not sustain men or vegetation because of the excessive heat of the sun. On the opposite side, the arctic and antarctic circles, being adjacent to each other, are uninhabitable because they are far removed from the sun. His juxtaposition of arctic and antarctic circles and his accompanying diagram indicate that he was conceiving here of five circular areas on the earth's flat surface.

In assigning two books of his *Etymologies* to physical geography, Isidore established himself as the great authority on the subject in Western Europe during the Middle Ages. For the next few centuries he was the sole source or one of the main sources of virtually all the geographical writing in the West. Whenever we handle a Latin geographical treatise, early or late, we are reminded that geography was not one of the more highly developed fields of classical scientists. Navigators in antiquity had serviceable pilot guides and, after Agrippa's survey, important officials traveling overland had access to itineraries; but when Latin geographical writers combined the two, tacking names of interior regions onto a coastal scheme, the result was as bad as could have been expected. Pliny's influence upon geographical writers in the West is evident for more than a thousand years. His mention of thousands of place names becomes hundreds and then tens in succeeding generations. Isidore's geographical books are largely made up of paraphrases and excerpts, approximately 150 in number, from Solinus' *Collectanea*. To Isidore's credit, it should be pointed out, is the fact that his account is not so much a bestiary or collection of curiosities as is the work of Solinus.

Pliny tried to offset the tedium of extended lists of the place names of larger inland areas by focusing disproportionate attention upon such conspicuous features as rivers, mountains, lakes, fountains, and supernatural or strange local phenomena, and by varying his over-all scheme, first taking up coastal and inland places and then returning to the adjacent islands, sometimes mere rocks. In the geographical treatises of later compilers Pliny's extended lists almost disappeared, while his conspicuous features took on greater prominence, becoming entities or categories detached from regional geography. Isidore has separate chapters in Book XIII on springs and lakes of the world with therapeutic or magical properties, on the ocean, on the Mediterranean Sea—the first recorded use of the word as a proper name—on gulfs of the world, on tides and straits, on lakes and pools, on the great abyss of waters under the earth, on the rivers of the world, and on biblical and classical floods.

Book XIV is devoted to regional geography. The bulk of the material is Plinian, derived from Solinus, but the arrangement is strikingly different. Pliny began with the Straits of Gibraltar, adhering to a common practice of periplus geographers. Solinus, possibly following Varro's precedent, began with Rome and Italy, went eastward, and picked up the countries of Western Europe at the end of his northern periplus. Isidore assigns single chapters of extended length to each of the three continents and—surprisingly, for those who are accustomed to classical orientations—he gives first position and the greatest amount of attention to Asia. In this respect his scheme conforms with the attitudes of Christian geographers and cartographers. They oriented their maps with east at the top and gave a prominent spot near the upper margin to Paradise and another, beneath it, to Jerusalem. Circular maps with Asia occupying the large upper segment and Europe and Africa the lower left- and right-hand compartments, with lines drawn in the form of a T to separate the continents, are known as T–O maps. They are among the commonest types in medieval cartography.[4] Such a map accompanied Isidore's text.

Isidore's chapter on Asia consists mainly of paraphrases and excerpts from Solinus, augmented by extracts from Christian writers on geography. Orosius' *Histories*—a work on universal history (c. 418) which was prefaced by a brief conspectus of the geography of the known world—was one of his main Christian sources. Then follow extended chapters on Europe, Africa, and the islands of the world, and short chapters on promontories and mountains. The summary of the geography of Africa includes a vague statement about a fourth region of the world, lying to the south, on the other side of ocean, cut off from the northern hemisphere by the sun's heat. Isidore cautiously remarks that reports of antipodean inhabitants are fabulous. The first chapter of Book XV deals with the founding and names of the cities of the world, with particular emphasis upon the Holy Land.

Tracing Isidore's influence upon later medieval writers would require a stout volume. His *Etymologies* became, for instance, one of the most important works in the history of glossography. In its range of subjects the *Etymologies* more nearly approximates our ideas of an encyclopedia than does any other classical work. It supplanted Pliny's *Natural History* as the most popular reference work of the Middle Ages, furnishing readers with the ultimate desideratum of a single book containing handily classified and intelligible information on all subjects. In providing a ready compendium of all knowledge which replaced other works of secular learning, Isidore has been thought to have done more harm than good. It is not likely, however, that students in the centuries immediately following Isidore would have been better educated without his books.

The high esteem enjoyed by Isidore's laboriously extracted compilations testifies to the aridity and stagnation of secular learning on the Continent during the seventh century. In the meantime, vigorous intellectual developments had been taking place in Ireland and Britain, stimulated by the missionary zeal of educated monks and by heated religious controversies between the Roman Church and its Celtic branch. By the early seventh century several monasteries in Ireland were gaining a reputation for secular as well as theological learning. Englishmen of high and low rank, according to Bede, began crossing the Irish Sea seeking knowledge and religious calm. At the same time England became the principal mission field of Irish monks, who crossed to western Britain or came down from Scotland. Meanwhile, in 597 Gregory I had sent a missionary party under Augustine from Rome to the east coast of Britain. Within six years Augustine had converted Kent and a part of Essex. As the Irish monks fanned out from northern and western borders and the Augustinian parties from Essex and Kent, the rival adherents confronted each other in many localities. Their common devotion to conversion of the heathen was not strong enough to overcome their bitter differences about matters of observance and, in particular, the method of reckoning the date for Easter. Augustine made two futile attempts to get Celtic missionaries to adopt Roman practices.

Paulinus, Augustine's colleague, was quite effective in Christianizing Northumbria, whose king, Edwin, was married to a Christian princess; but upon Edwin's death Paulinus was driven from the land. Under the next king, Oswald, who had been educated at Iona, an Irish monastery off the coast of Scotland, the conversion of Northumbria was undertaken anew, this time by Irish missionaries. Oswald was succeeded by his brother Oswy, who greatly expanded his domains and espoused the cause of Christianity. After extensive conversions in Mercia and Essex, Oswy decided to summon representatives of the Roman and Celtic churches to a synod at Whitby (664) to settle religious differences;

upon hearing both sides, he adopted the practices of the Roman Church. Northumbria and Kent were thus brought into close union, and the two kings applied for a new archbishop to fill the vacant see at Canterbury. Pope Vitalian appointed a Greek monk, Theodore, an outstanding scholar, hoping thus to strengthen Roman partisans in controversial matters. Theodore, who had been educated at Athens, and his colleague Hadrian, formerly an abbot of a monastery near Monte Cassino, brought an atmosphere of Greco-Roman erudition to England. Wilfrith, the leading English supporter of the Roman Church, visited Rome three times; and Benedict Biscop, who became abbot of St. Peter's at Canterbury in 669, went there at least six times. Biscop happened to be in Rome when Theodore was appointed archbishop of Canterbury, and he returned to England in 669 as Theodore's guide and interpreter. On his next visit to Rome, in 671, Biscop acquired by gift and purchase a sizable collection of Greek and Latin manuscripts.

Much attention has been drawn to Theodore's introduction into Britain of Greek secular and theological learning, and importance has been attached to Bede's account of Theodore and Hadrian gathering about them a band of disciples, instructing them in the arts of metrics, astronomy, and computus, as well as religion, and enabling them and their pupils in turn, even to Bede's day, to speak Latin and Greek as fluently as their vernacular. Bede appears to be writing with a hagiographer's enthusiasm. Elsewhere he speaks of Latin, Greek, and Hebrew as the three sacred languages. But there is little tangible evidence of fluency in Greek in the writings of those English Hellenists; and, as the studies of Laistner and others have revealed, Bede himself, though he was adept at handling Greek, made very little use of Greek sources and preferred to use Latin translations, if they were available.[5] Bede's classical learning was derived more largely from Irish sources than through Augustine, Theodore, and Hadrian.

When King Oswy died in 671, his son Egfrid succeeded him on the throne of Northumbria. Egfrid gave Benedict Biscop a grant of land to build a monastery in honor of St. Peter on the north bank of the Wear River. The foundations of Wearmouth were laid in 674. Egfrid was pleased with the result and, about seven years later, he gave Biscop a second grant to found another monastery, in honor of St. Paul, a short distance away at Jarrow, on the River Tyne.

The infant Bede was born in 672 or 673 on the very land that Egfrid shortly before had deeded to Biscop for a monastery. The child might almost be said to have been assigned with the land to the Lord's service because, at the age of seven, he was delivered by his kinsmen into the care of Benedict Biscop, now abbot of Wearmouth, with the promise that he would devote his life to the monastery. When the buildings at Jarrow were completed to form a double monastery with Wearmouth, Ceolfrid, prior at Wearmouth, was appointed abbot at Jarrow; and seventeen monks were transferred to the new quarters, the young lad Bede probably among them. Bede's life was one of complete dedication to Christian service. His vast erudition nearly always bore some connection to religious matters. Biscop's collection of manuscripts formed an excellent foundation for a library at Wearmouth-Jarrow, and Ceolfrid almost doubled the size of the collection. Bede began his study of Latin upon his arrival at the monastery. His more advanced studies prepared him to read Greek authors and to calculate, through application of his knowledge of mathematics and astronomy, the dates for movable feasts. Bede was the best-read scholar of his age in the West, and during his lifetime Jarrow became the foremost educational and literary center of England.

Bede ranks with Boethius at or near the top among intellectual figures of the Latin West during the early Middle Ages. Both men had extraordinary mental endowments and both were stimulated by their environment, Boethius by the challenge of the pitiable state of learning in his day and the opportunities afforded by Theodoric to introduce Greek philosophy into Italy, Bede by the intensity of religious controversy between the rival churches in England—the most acute question, the dating of Easter, calling for some knowledge and skill in mathematics and astronomy.

Bede's literary productions fall into three categories: scientific, historical, and theological. His early writings, like Boethius', were of the textbook variety, designed for school use. His later writings show some well-developed critical faculties, on occasion even an interest in the observation of natural phenomena, and, by medieval standards, a gentleman's scruples about acknowledging his sources. Judged by his greatest work, *The Ecclesiastical History of the English Nation*, he could not be called a critical historian, but he does have regard for the worth of his sources and the difference between presumed facts and casual reports. He shows the almost unique characteristics, for a medieval man, of understanding what he is appropriating or assimilating from earlier authorities and of striving to shape his borrowed materials into a consistent whole.

All of Bede's writings were intended to propagate the Christian point of view. Unlike Aldhelm and the coterie of Hellenist disciples of Theodore and Hadrian, he did not have much interest in classical literature. Laistner has pointed out that Bede's quotations of single lines from Terence, Lucilius, Varro, Lucretius, Sallust, Horace, Ovid, and Martial do not indicate that he was familiar with their works; the same quotations are found in Isidore's works and in the Latin grammatical treatises which were in Bede's library. Virgil is the only classical poet with whom Bede can be said with assurance to have been familiar. The other classical work that was certainly available to him was a fragmentary copy of Pliny's *Natural History*. Some scholars have assumed that Bede was using an abridgment of Pliny or a collection of Plinian excerpts of the sort that were bound with medieval tables for calculating movable dates. But Pliny's work was more widely circulated in fragmentary form than in entirety. Entire chapters of Bede's *On the Nature of Things* were copied verbatim from it; his borrowings may be traced to Books II, IV, V, VI, XII, XVI, and possibly XXXVII. Bede's copy may have consisted of the first half of the *Natural History*; it certainly did not include Book XVIII.

On one occasion when Bede has appropriated a whole chapter from Pliny's text, he acknowledges the fact and recommends that his readers consult Pliny for further information. Considering Bede's scruples about acknowledging some of his indebtedness to Pliny, it is curious that he is not so considerate of his other major source of scientific information, Isidore. He does not seem to have had much respect for Isidore's authority, yet he appropriates material from his *Etymologies* and *On the Nature of Things* as readily as from Pliny, and only thrice reveals his source, each time merely to take issue with Isidore. This practice was used by several postclassical Latin compilers when they wished to impress readers with their learning and technical competence; but Bede does not display the mannerisms of the run-of-the-mill compiler. Why did he show a particular animosity against Isidore? Was it perhaps because he was aware, from the common sources they both used, especially in theological writings, that Isidore had seldom made true acknowledgment of his pilferings, and was this Bede's way of indicating his disapproval?

Bede's early treatises on metrics, figures of speech, and orthography, though they were wholly unoriginal compilations, did not follow the conventional style of manuals on trivium subjects. These books were designed for the particular needs of the monks at Jarrow and other English monasteries. His style was concise and simple and his approach direct and practical, free from the irrelevant and outmoded discussions that characterized most grammatical treatises in his day. His two elementary scientific textbooks, *On the Divisions of Time* and *On the Nature of Things*, were intended to provide students with the necessary background and technical information for computing the dates of movable feasts. *On the Divisions of Time*, written in the year 703, does address itself directly to calendar problems, but *On the Nature of Things*, is a rather conventional primer of Christian cosmography and the principles of astronomy. Its date is not known, but it undoubtedly was an early work, probably written in conjunction with *On the Divisions of Time*. On three occasions Bede refers to them together. More importantly, an examination of the tables of contents of Bede's treatises reveals that the opening chapters of *On the Divisions of Time* follow the order and subject matter of Isidore's *On the Nature of Things*, and the remainder of the latter provides the framework for Bede's entire treatise *On the Nature of Things*.

Bede was more intelligent and more expert in handling his materials than Isidore. It has been suggested that Bede's advantage over Isidore was that he had access to a large portion of Pliny's encyclopedia; but this would have been a mixed blessing. Pliny did furnish Bede with the conception of the earth as a sphere and with some information about planetary anomalies not found in Isidore; however, he was also responsible for two of Bede's grosser misconceptions—and, as it happened, Isidore, who was following Cassiodorus on these matters, was better informed. The superiority of Bede's treatises should rather be attributed to the superiority of his mind over Isidore's. Bede's scientific library was a restricted one, probably much more so than Isidore's. Aside from his Irish computistical collections, it appears that he had to depend upon Isidore and Pliny. He did not possess copies of the widely circulated works by Chalcidius, Capella, and Boethius. His Christian cosmography was largely derived from Augustine, Ambrose, Basil, and Gregory the Great.

Bede's *On the Nature of Things* is a concise textbook of fifty-one chapters, superimposing a Christian cosmography upon a framework of secular cosmographical and astronomical traditions. The waters above the firmament now become the demarcation between the corporeal and incorporeal realms. Isidore's ambiguous statement about the seven heavens of the seven planets, traced by him to Ambrose' association of Platonic concepts of the harmony of the spheres with a verse in Psalm 148 ("Praise Him, ye heavens of heavens"), takes on a new aspect. There are now seven distinct heavens about the earth: air, ether, Olympus, the fiery realm, the stellar firmament, the heaven of angels, and the heaven of the Trinity.

Bede's description of the corporeal universe is mostly a conflation of extracts from Isidore and Pliny, some verbatim, the greater number recast, some mere phrases, others verbatim transcriptions extending over an entire chapter. The chapter titles and arrangement of material closely follow Isidore's treatise. The corporeal universe consists of sky and earth, made of the four elements shaped to a spherical form. The position of the elements is determined by their weight, the heaviest, earth, being at the center and bottom. The similarities and differences of the elements link them in a stable union, a venerable and oft-repeated *Timaeus* doctrine. Bede's chapter on the five celestial parallels indicates that he had Isidore's text before him, but he does not fall into Isidore's blunder of confounding celestial circles and terrestrial zones. Rather he follows Pliny, extracting parts of his text verbatim. He copies Pliny's statements about the supposed influence that stars have upon the weather but points out that they have no influence upon man. He repeats Isidore's explanation that the retrogradations of the planets are caused by the powerful effects of the sun's rays. Bede's entire chapter (14) on the anomalies of planetary motions is copied from Pliny, this time with full credit. He next copies Pliny on the colors of the planets, and takes the following chapter on the zodiac verbatim from Pliny. Bede shows discrimination whenever he handles Pliny, and he here omits Pliny's nonsensical explanation that animals are born in deserted parts of the globe because Venus occasionally transgresses the limits of the zodiac. Again Bede had Isidore's text before him for the discussion of the sizes of the sun and moon, but this time he makes the mistake of accepting Pliny's statement that the moon is larger than the earth. Isidore had the correct view of the sun as the largest and the moon as smallest of the three, obtaining his information from Cassiodorus, who was using some Ptolemaic authority here. Bede got his information about eclipses from Pliny, using brief extracts in *On the Nature of Things* and, in a larger work, *On the Reckoning of Time*, copying almost his entire chapter from Pliny, with acknowledgment. Bede's discussion of the differences in time of the observations of eclipses in the east and west was also derived from Pliny.

For meteorological and terrestrial phenomena Bede also depended upon Pliny and Isidore, following Isidore's titles and arrangements of subjects very closely. Thunder is caused by the bursting of clouds, lightning by their clashing together. Comets portend the fall of kings, pestilence, and other dire events. Pliny is copied on the longest and shortest known periods of comets. The salt in the seas is heavy and cannot be lifted by the sun's rays—the same rays, presumably, that elsewhere could stop planets in their courses. Consequently rain, rivers, and lakes are not salty. Earthquakes result from violent winds rushing through cavernous spaces, and Aetna's eruptions are caused by sea water coming through subterranean channels. Bede foolishly adopts the seven absurd geographical parallels which Pliny was laying down merely to parade lists of exotic place names. Pliny's parallels in the east ranged from the southern extremity of India to the latitudes of central Russia, and converged in the west upon the Iberian peninsula. Bede reduces Pliny's lists of names and adds Pliny's three mathematical parallels in the north —Don and Dnieper mouths—Gaulish coast; Hyperboreans-Britain; and Rhipaean Mountains—Thule. He adds two other mathematical parallels, through Meroë and Syene, which he probably got from Isidore.

Like other Churchmen of the Middle Ages, Bede was interested in scientific knowledge only as it had application to a devout Christian life. The problem of ascertaining the correct date for Easter and of establishing a uniform observance had long been a grievous one throughout the Christian world, particularly in borderlands remote from the ecclesiastical bodies that made such authoritative decisions. Northumbria became a battleground for the Paschal question. The monks of Iona promulgated the Irish system of dating, while the less numerous but more potent representatives of the Kentish Church defended the Roman system. At one time King Oswy was fasting for Lent while the queen was feasting for the Resurrection. Oswy presumably had settled the matter at Whitby in 664 by deciding in favor of Roman practices, but the Northumbrian Church had predominantly Irish affiliations, and Irish observances persisted despite the decree. Computists in Northumbria were depending for guidance upon Books V and VI of Isidore's *Etymologies* and an Irish computus of the seventh century. Bede, with his low regard for Isidore's competence in such matters, decided to try to reach a definitive solution by his own painstaking studies. His larger work on the subject, *On The Reckoning of Time*, according to Poole [6] established his reputation as the greatest master of chronology in the Middle Ages.

The Easter problem was essentially one of adjusting a Hebrew-Christian celebration, based upon a lunar calendar, to the Julian (solar) year—the very sort of problem that had vexed the ancient Greeks from earliest times, since they, like the Hebrews, based their calendars upon lunations. At some unknown date, in central Greece perhaps as early as the Mycenaean Age, the Greeks had adopted an 8-year solar cycle, with 3 intercalated lunar months during each year. By the fourth century B.C. a more precise system, known as the "Metonic" cycle, perhaps a Babylonian invention, was in use in Athens. This was a luni-solar cycle of 19 years, with 7 intercalated months. The average length of a Metonic year was $365\frac{5}{19}$ days. Discrepancies in Easter dating had vexed the Church councils since the second century. It is not a little surprising to find that chronology experts in the Church, including many from eastern branches who had been educated in Greek schools and were able to avail themselves of Greek scientific acumen in calculations and astronomical observations, had to arrive by trial and error at results that the Greeks had reached nearly a millennium earlier. Computists in the Roman Church for a long time experimented with an 8-year cycle, and it was tried in various forms in the churches of the East; 11-, 16-, and 84-year cycles were also widely adopted, and centuries had to pass before the superiority of a 19-year cycle was recognized.

Bede's first treatise on the calendar problem, *On the Divisions of Time (De temporibus)*, is an extremely concise manual of sixteen chapters in only nine pages, but it was prepared with much care. When a revision was undertaken in 725, there was little in the original work that needed correction. *On the Divisions of Time* addresses itself immediately to the subject of the title, beginning with the smallest divisions—minutes and hours—and proceeding to days, nights, weeks, and months, a chapter for each. This portion of his treatise was reproduced, with a sense of discrimination, from the opening chapters of Isidore's *On the Nature of Things* and from his *Etymologies* Book V. Chapter 6, dealing with the Roman months, consists largely of paraphrases and extracts from Macrobius' *Saturnalia* I.12–15 and Isidore's *Etymologies* V and *On the Nature of Things* 4. It appears that the only portion of Macrobius' work known to Bede was this fragment of four chapters of Book I. Bede's chapter 7, on the solstice and equinox, was nearly entirely extracted verbatim from Book II of Pliny's *Natural History*. Chapter 8, on the seasons, was largely extracted from Isidore's *Etymologies* V and *On the Nature of Things* 7. Chapter 9, on the various kinds of years, was obtained mainly from Isidore's *On the Nature of Things* 6. Bede then proceeds to show his readers how to calculate Easter tables from a 19-year cycle. Little is known about the sources of his later chapters. The borrowings in his early chapters were unacknowledged but have been traced.[7]

The monks at Wearmouth-Jarrow pleaded with Bede for a long time to prepare a new version of *On the Divisions of Time* because its extreme conciseness made it difficult to follow. In 725 he responded with a new book, based upon the old one but expanded twelve times in size— so that the nine pages became over a hundred. *On the Reckoning of Time (De temporum ratione)* is Bede's masterwork in science. His early treatises, though they give evidence of critical ability to select and reject material derived from earlier authorities, were in other respects typical compilations. If the direct quotations, nearly all unacknowledged, were deleted from these manuals, there would be little left. *On the Reckoning of Time*, by contrast, reveals a mature scholar and thinker. Bede has thoroughly assimilated his materials and is ready to explain patiently and familiarly to his monks the intricacies of calendar reckoning. In several places we find him expatiating in his own words with very little, if any, traceable dependence upon authorities. The book has a friendly approach, not the cold, mechanical style of a compilation. There is a table for beginners in a chapter entitled "For Those Who Cannot Compute the Day of the Moon." Other chapters are entitled "Why Does the Moon Have Prone, Supine, and Upright Aspects?" and "Why Is It That the Moon, Though It Has a Lower Position, at Times Appears Higher Than the Sun?" We seem to be browsing in a medieval Plutarch. Bede and Plutarch both derived their science entirely from books and both succeeded in giving readers the impression that they understood everything they were discussing; on two occasions Bede uses a device dear to Plutarch of introducing an observation from nature as an example. To illustrate the optical illusion in the apparent positions of the sun and moon he calls attention to lamps suspended from the ceiling of a cathedral, as seen by those below.

Occasionally Bede reverts to the habits of a compiler. Two chapters of *On the Reckoning of Time* were copied from Macrobius, one from Pliny, and one from Ambrose and Basil; and there are extended quotations from Pliny, Augustine, Macrobius, and pseudo-Hippocrates. The striking difference is that now, with one exception, Bede acknowledges the sources of longer extracts, usually by book and title as well as author. As C. W. Jones has pointed out, Bede in this work was relying mainly upon Irish computus sources, and makes frequent use of patristic commentaries and biblical citations.

On the Reckoning of Time contains three features that are worthy of note. Chapter 1 presents the main and almost unique account of early finger reckoning, a practice which was popular in the ancient world; and chapter 65 provides Easter tables for the years 532 to 1063. Chapter 29 contains the best extant account of tidal motions that had yet been written. Bede notes the differences between spring and neap tides and the effect of winds in retarding or advancing the flow of a tide. He writes as if from personal observation about the differences in times of tides at different points along the same shore. This phenomenon must have been common knowledge among fishermen and mariners, but, in any case, Bede's statement is the earliest record of the important principle of the "establishment of a port."[8] On another occasion Bede showed a typical compiler's trait when he wrote at length about the structure of the Roman Wall only ten miles from Jarrow, while depending for his information upon Vegetius and Orosius and not taking the trouble himself to examine the wall.[9]

With *On the Reckoning of Time*, as has been said, Bede took his place as the great medieval authority on chronology in the Latin world. The system of B.C. and A.D. dating he used was derived from Dionysius Exiguus, but it was Bede who made the practice popular. As we shall see in the next chapter, Bede became the key figure in the Carolingian Renaissance; and his fondness for Pliny's *Natural History*—"that most excellent [*pulcherrimum*] work," he called it—was also responsible for the revival of interest in Pliny during the Carolingian Age.

XIV

ROMAN SURVIVALS IN THE LATER MIDDLE AGES

The Continent was next after England to feel the quickening effects of the scholarly awakening that had begun in Ireland. Charlemagne realized, early in his reign, that drastic reforms would have to be introduced into the educational system of his kingdom in order to alleviate the abysmal conditions of ignorance among clergy and civil servants. He invited several Italian scholars to his Palace School at Aachen and made one of them headmaster; but the results were not impressive. In 782 he persuaded Alcuin, headmaster of the cathedral school at York, to accept the headship of the school. Under Alcuin, York had gained precedence over Jarrow as the foremost educational center of England. Alcuin introduced at Aachen an English system, strongly influenced by Bede's teachings and writings. Charlemagne himself attended classes at the school and, with the help of Alcuin, he undertook the reorganization of schools throughout the kingdom. The combined efforts of Church and state produced Europe's first regularized educational system. The curriculum at the Palace School was broadened to include the liberal arts, and students outside the court were admitted. The emphasis was of course upon religious teachings, but Charlemagne's secular needs had to be considered also.

The textbooks that Alcuin composed for use at the school are exceedingly elementary, all but one being in dialogue form to make them attractive to pupils. These manuals give little indication of an outstanding teacher and of a school that was to produce the leading schoolmasters of the next generation. The simplicity of the dialogues strikes us as amusing. Alcuin used the standard authorities for his textbooks on rhetoric, grammar, orthography, and dialectic. His *Propositions for Sharpening Youthful Minds* contains fifty-three questions, some requiring skill in calculation, others mathematical ingenuity. The best known is the problem of three men and their three sisters having to cross a river in a boat holding only two persons, assuming that each girl, to be safe, must have no other companion than her brother. Such exercises prepared a student to proceed to a study of the computus literature, which was also imported from England and Ireland.

Scientific knowledge was at a low ebb during the Carolingian Age. Ludwig Traube was one of the first scholars to realize the extent of the deceptions practiced by medieval writers who were trying to give their readers the impression they were familiar with Greek.[1] The Greek words and phrases used by Alcuin and Hrabanus (or Rabanus) Maurus, as he pointed out, they had derived from the works of St. Jerome. Bede, Isidore, Pliny, Cassiodorus, Solinus, and Capella were the popular Carolingian authorities on science. The Reichenau and Murbach library catalogs of the ninth century also list several copies of Germanicus' translation of Aratus' *Phaenomena* and Hyginus' *Astronomicon*. The popularity of Bede's calendar treatises is demonstrated by the number of extant manuscripts, over 130 of *On the Reckoning of Time* and over 60 of *On the Divisions of Time*. Such was the reputation of the Venerable Bede that four or five centuries after his death spurious treatises were still being attributed to him.

Hrabanus Maurus, a deacon at Fulda, was sent to Tours to complete his education under Alcuin, during the last years of that scholar's life. Hrabanus became one of the outstanding teachers and scholars of the age, with a reputation surpassing that of his master. Of his abundant writings, many of which are unpublished, three are of interest to us. *On the Education of Clerics* is a compilation derived chiefly from Isidore's *Etymologies*, Cassiodorus' *On Profane Literature*, Augustine's

On Christian Doctrine, Gregory's *Morals*, and Bede's works. Boethian arithmetic, already reduced by Cassiodorus and Isidore, became so attenuated in Hrabanus' book that very little is left which would serve to instruct students. Another treatise, on the calendar *(De computo)* was derived almost wholly from Bede's *On the Reckoning of Time*, although the author makes no mention of Bede. A voluminous work entitled *On the Universe* or *On the Nature of Things* is little more than a rearrangement of Isidore's encyclopedia, consisting of 22 books, two more than Isidore's, and occupying 600 columns in Migne's *Patrologia latina*. It omits the first four books, half of Book V, and the tenth book of Isidore's *Etymologies*, copies the rest, beginning with Book VII verbatim, omitting a great deal and adding extracts from Augustine, Alcuin, and others. By his transpositions, Hrabanus was able to begin with theological materials. He could conveniently omit Isidore's books on the liberal arts since he had already handled these in *On the Education of Clerics*. About the sphericity of the earth Hrabanus thought it best to say nothing, but he does try to reconcile the circular horizon with the scriptural four corners of the earth. He feels that Euclid's inscribing of a square within a circle will perhaps obviate the difficulty.

The state of geographical knowledge in the Carolingian Age is indicated by two short treatises, a *Survey of the World*, written by an Irish monastic scholar named Dicuil in 825, and an anonymous *De situ orbis*, written shortly after 850. Christian authors had opportunities to gather firsthand information about many regions that were virtually unknown to writers at the height of the Roman Empire. An episcopal see is believed to have been established at Samarkand, for instance, in the fifth century. Christian monks and missionaries traveled across much of the known world and had good means of communicating with each other. Under such conditions it is disappointing to find that most writers were satisfied, as was Isidore, to extract and garble information recorded by Pliny and Solinus, some of it dating from Roman Republican times.

The brief *De situ orbis* is of little interest, being a compilation of extracts from Mela, Solinus, and Capella. Dicuil's treatise is a typical geographical compilation of the period. A large amount of it was extracted verbatim from Pliny and Solinus. Since Dicuil was a careful scribe with a particular interest in the accuracy of textual readings, it is sometimes possible, as Mommsen pointed out,[2] to emend the texts of Pliny and Solinus by comparing them with Dicuil's text. Dicuil's later years were spent in an Irish monastery located somewhere in the Carolingian state. We are naturally disappointed to find that this first geographer of the Frankish kingdom tells us nothing about Saxony and Bavaria and depends for his geography of Germany upon Pliny, who, as has been remarked, was no better informed and, in some places, less well informed than Herodotus. On the favorable side, Dicuil does record some of his intimate knowledge of islands on the northwest coast of Britain in the vicinity of his birthplace and reports two interesting journeys. The first is an account of a six-months visit to Iceland (795) by Irish missionaries, who described the midnight sun and other celestial phenomena and the ice conditions in the sea north of the island. The second account tells of a pilgrimage to Jerusalem undertaken by an Irish monk called "brother Fidelis." Sailing along the Nile he passed the pyramids of Giza, which he took to be granaries built by the biblical Joseph during the years of plenty. This notion seems to have been substantiated during the Middle Ages by the popular derivation of the word "pyramid" from the Greek word for "wheat" (*pyros*). From the Nile he sailed into the Red Sea by way of a canal which was finally closed in 767. Beazley dates the pilgrimage about 750–760.[3]

Another of the many Irish scholars to immigrate to Frankland during the Carolingian Age was John Scot Erigena (c. 810–877). Nothing is known about Erigena's life in his home country. His ability to read Greek commended him to Charles the Bald, who commissioned him (c. 855) to revise a recent translation of the *Corpus Dionysiacum*, popularly ascribed to Dionysius the Areopagite or to St. Denys. He then translated other Greek theological works, some at the instance of the king. Erigena's significance in intellectual developments of the West lies chiefly in his revival of interest in Platonism. He became the leading exponent of Neoplatonic cosmography for the next three or four centuries, by which time the scholars of Chartres had acquired a more intimate knowledge of Neoplatonic doctrines. Despite his ability to read Greek, Erigena derived his instruction not from reading the Neoplatonic masters—Plotinus, Porphyry, and Proclus—but secondarily, from Chalcidius' translation and commentary on the *Timaeus* and from Capella's encyclopedia and Macrobius' *Commentary*. Duhem is naive here, as he is on numerous occasions, in assuming that because Erigena makes no reference to Macrobius' work he was not influenced by that important Neoplatonic writer. Several direct borrowings have since been pointed out.[4]

Contrary to common suppositions, Erigena had little competence in handling scientific subjects. In *On the Divisions of the Natural World* he offers a pretentious account of Eratosthenes' methods used in obtaining a figure of 252,000 stades for the earth's circumference; he or his intermediary expanded this explanation from Capella's faulty account. He goes on to explain that, if you divide Eratosthenes' figure for the circumference by two, you get a measurement of 126,000 stades for the earth's diameter. A little later he repeats the assertion that the figure for the earth's diameter is half that of its circumference and observes that the measurement of the earth's diameter is equal to that of the moon's distance.[5] In the same work Erigena discusses the variations in the color of the planets as they traverse upper and lower regions of the heavens. Saturn, close to the fixed stars, is cold and pale. Jupiter, Mars, Venus, and Mercury, however, circle about the sun, "as Plato teaches in the *Timaeus*." When they are above the sun they appear clear, but below the sun they are red. There is no reason to regard this statement as a brilliant anticipation of the Tychonic system, as Duhem does.[6] It seems that Erigena, or his intermediary, was intrigued by the color variations of planets and, because of this interest, expanded Chalcidius' or Capella's Heraclidean doctrines about the heliocentric motions of Mercury and Venus to include two more planets whose differences in brilliance are quite noticeable. In attributing these views to Plato, Erigena was indulging in a thousand-year-old foible.

A lesser known figure of the Carolingian Age was Helpericus, a monk who lived in the middle of the ninth century and taught grammar and the computing art at Auxerre. His *Computus*, in thirty-seven chapters, was designed for beginners, among whom he graciously counted himself. His work was derived mainly from Bede's *On the Reckoning of Time*, with minor borrowings from Macrobius' *Commentary*. Helpericus' treatise enjoyed a vogue, alongside Bede's work, until the thirteenth century, another indication of the keen interest that clerics had in trying to determine the correct date of Easter. The author of the *Computus* is not to be confused with another Helpericus who lived a century later and whose manuscript treatise *In calculatoria arte* was studied with care by Duhem.[7] The later Helpericus was an avowed disciple of Macrobius, and his treatise reveals many borrowings from Macrobius' *Commentary*. However, at one point he undertakes to argue against those who suppose that the planets move in a direction contrary to that of the celestial sphere. Even the most palpable and incontrovertible celestial phenomena were not immune to refutation from medieval writers who fancied they were displaying their acumen and independence.

The *Liber de mundi constitutione*, falsely ascribed to Bede but probably post-Carolingian in date, is another treatise that received considerable attention from Duhem.[8] Despite the fact, or perhaps we should say because of the fact, that the anonymous author cites several of Bede's works, some inattentive medieval scholars attributed the treatise to Bede. Duhem may have been correct in assigning a date later than Erigena to this treatise, but to argue it on the basis of Erigena's unfamiliarity with Macrobius has no bearing, since Duhem was mistaken in supposing that Erigena was not influenced by Macrobius. Duhem's careful examination of the contents of the anonymous treatise reveals its heavy dependence upon Bede and some indebtedness to Macrobius. At one point almost an entire chapter of Macrobius' *Commentary* (I.18) was copied verbatim.

The first glimmer of the dawn of genuine Greek science that was to rise upon Western Europe through the Arabs is found in the writings of Gerbert—although it must be supposed that Latin scholars before his time had already come in contact with Arabic translations. The dawn was to come very gradually, as men with real scientific instincts were to appreciate the superiority of Greco-Arabic science over the handbook science of the Latin West. Latin science continued to find favor even beyond the Renaissance into modern times, but in the measure that the new science came to be appreciated and cultivated, Latin handbook science generally fell into neglect and suffered a decline. However, there was always an advantage working in favor of the venerable Latin texts: Latin was the established and familiar language of learning—and science continued to be, in the minds of most intellectuals, merely another branch of learning.

Gerbert was born in Auvergne about 940, received some of his education in the County of Barcelona, began teaching at Reims in 972, and was Pope Sylvester II from 999 until his death in 1003. While in Barcelona he became absorbed in Arabic mathematical and astronomical studies and established several contacts with men of learning and science. He learned about Hindu-Arabic numerals and is the first man who is known to have taught the new set of numbers in the West. Gerbert did not study Arabic works in the original but used Latin translations, which were being produced since the ninth century. In 984 he addressed a letter from Reims to Bishop Lupitus (Lopez) of Barcelona, requesting him to send a copy of a *Liber de astrologia* that Lopez had translated from Arabic.[9] Bubnov discovered in the Bibliotheque Nationale a fragment of a treatise on the astrolabe, translated and adapted from Arabic sources. Duhem believes that the author of an extant treatise, *Liber de astrolabio*, used this very fragment in composing his work, and surmises that it may be a fragment of the treatise that Gerbert requested from Lopez. The attribution of the *Liber de astrolabio* to Gerbert is not positive but Duhem regards it as fairly certain.[10] Gerbert wrote another treatise, *On Reckoning with the Abacus*; this, too, was derived from Arabic sources.

Gerbert overawed his contemporaries with the astronomical instruments he had acquired from Spain and with his knowledge of mathematics and science. Duhem sees Gerbert, in his use of Arabic sources for his book on the astrolabe, as introducing a new vogue.[11] During the eleventh century several Latin treatises on astronomical instruments appeared, closely following Arabic models. A notable imitator of Gerbert, according to Duhem, was Hermann of Reichenau (1013–54), who wrote on the astrolabe and abacus. Mention should be made, in passing, of the treatment given to the astrolabe by Hugh of St. Victor (d. 1141). He discusses its uses in the opening chapters of the practical part of his *Practica geometriae* and in chapter 42 suggests the possibility of determining the sun's distance by observations from two widely separated points—a Greco-Arabic procedure. But he goes on to include Macrobius' puerile discussion of the dimensions of the sun's orbit and of the relative sizes of the earth and sun.

Gerbert was also famous for his library acquisitions. He has been called the greatest book collector of the Middle Ages. His collection included a copy of Boethius' translation of Nicomachus' *Introduction to Arithmetic* and the so-called "Geometry of Boethius," an eleventh-century compilation which only faintly reflected Boethius' lost work. Gerbert's book on geometry shows familiarity with Boethius' translation of Nicomachus' *Introduction*, but it is so different from Euclid's *Elements* that he could not have handled that work either in its Boethian form or in any other faithful adaptation or translation. Interest in geometry was to continue to be slight for the next century or two, with only practical applications of some Euclidean propositions. Euclidean conclusions served the purposes of surveyors and engineers who could not begin to demonstrate those conclusions.

Let us return for the moment to books in the Latin tradition. A short compendium of cosmography, astronomy, and geography, known by the title *De imagine mundi*, was formerly ascribed to St. Anselm or to Honorius of Autun; it is now more reasonably, though with some doubts, attributed to Honorius Inclusus, also called Solitarius. This treatise was compiled from classical Latin and early medieval sources. Honorius Inclusus was a Benedictine monk who appears to have flourished in England at the end of the eleventh century. The cosmographical portion of the work was derived from Pliny and Bede, with some traces of Macrobius' *Commentary*. The discussion of the apsides in planetary orbits was reproduced from Bede, who in turn was reproducing Pliny's statements. Bede was also the source of the scriptural conception of the upper realms of the universe contained in the work. The geographical material was obtained from Isidore. One scholar, however, has detected direct borrowings from Solinus' *Collectanea* in the section on the marvels of India.[12] *De imagine mundi* exerted considerable influence in the later Middle Ages. It provided the geographical framework of a popular book, *Imperial Leisures*, by Gervase of Tilbury, to which that author tacked on statements from Orosius, Isidore, and other authorities on geography. An anonymous treatise, *Image du monde* (c. 1246), got its title and most of its matter from the earlier work.

Manegold of Lautenbach, who became abbot of Marbach in 1103, reports in a tractate *Against Wolfhelm of Cologne* a lengthy conversation with his colleague of that name who was abbot of Brauweiler. He chides Wolfhelm for entertaining heretical doctrines, found in Macrobius' *Commentary*, about the four inhabited quarters of the globe, three of them permanently cut off from the known quarter by impassable belts of ocean and the scorching heat of the torrid zone. How can Wolfhelm reconcile the Saviour's death for all mankind with the notion that three-quarters of the earth's population can never learn about the Gospel? Manegold was himself quite interested in secular literature and wrote several commentaries—one of them, now lost, on Chalcidus' *Commentary on the Timaeus*.

The twelfth century marks the beginning of the spate of Latin translations of Arabic and Greek works that was to revolutionize intellectual life in Western Europe. Because of practical considerations scientific treatises received greater attention from translators than works of general literature and philosophy. At the beginning of the century Bede and Isidore were still the great authorities on science. By the end of the century many of the most important works of Greek and Arabic science had been translated into Latin, with some Latin translations of Greek originals competing with Latin translations of Arabic translations of the same Greek originals. Being the new language of science, Arabic enjoyed greater prestige during this period than Greek. As a consequence, Arabic versions of Greek works were often translated before the Greek originals.

A partial list of twelfth-century Latin translations will indicate the scope and extent of the scientific revival that was taking place. The following works were translated from Arabic versions of Greek originals: Euclid's *Elements* (at least three separate translations) and *Optics*; Apollonius' *Conic Sections* (Preface only); Archimedes' *Measurement of the Circle*; Ptolemy's *Almagest, Optics, Planisphere*, and *Quadripartitum*; various treatises of Hippocrates and his school, and of Galen; Aristotle's *Meteorologica* I–III, *Physica, De caelo*, and *De generatione et corruptione*; Theodosius' *Spherics*; and Alexander of Aphrodisias' *De motu et tempore*. Among works translated directly from Greek were Euclid's *Optics, Catoptrics*, and *Data*; Heron's *Pneumatics*; Ptolemy's *Almagest*; Hippocrates' *Aphorisms*; various treatises of Galen; Aristotle's *Posterior Analytics, Meteorologica IV, Physica, De generatione et corruptione, Parva naturalia, Metaphysica* I–IV, *De anima*; and Proclus' *De motu*. The trigonometric tables and *Algebra* of al-Khwarizmi were also translated into Latin, as well as various other Arabic works by Al-kindi, Thabit ibn Qurra, Rhazes, Avicenna, and lesser known writers.

While such scientific treasures were beginning to become available, some Latin writers on science seem to have remained completely unaware of the new developments; others who were aware of the new science found it hard to discard the congenial and ingrained doctrines of familiar authorities. In 1120 Lambert of St. Omer completed a large encyclopedic work, *Liber floridus*, which, as the title indicates, was an anthology culled from earlier sources, nearly all postclassical and early medieval. He complains in it that the thirst for knowledge has almost disappeared in his day. The sources of Lambert's extracts, most of them acknowledged, have been noted by Manitius. He devotes much attention to biblical scholarship, Church history, and world chronology, with extended treatment of the computus and calendar reckoning. Some of the liberal arts—grammar, orthography, mathematics, and astronomy—also get considerable attention. Lambert's authorities on cosmography and astronomy were Macrobius and Bede. His use of Isidore was much more extensive than is indicated in his citations. By this time the tags of Latin scientific doctrines had become commonplaces; it is therefore difficult, if not impossible, to determine who is to blame for scholarly remissness. Lambert's cosmographical views about the earth being a point in the exact center of the spherical universe and his figure for the earth's circumference (252,000 stades) came from Macrobius' *Commentary*. He also repeats from Macrobius the Cratean doctrines about four inhabited continents, transversely and diametrically opposed. The world maps accompanying Lambert's manuscripts are drawn in the Macrobian zone-map style, with symmetrical northern and southern continents separated by an equatorial ocean. One manuscript map has a large island on the southwestern margin, whose caption explains that antipodeans inhabit the island and have days and nights opposite those of northern inhabitants.[13] Lambert's doctrines about zodiacal constellations and the anomalies of planetary orbits were derived from Bede.

Another writer of this period to omit the new Arabic science and learning from his encyclopedic work was Hugh of St. Victor. His *Didascalicon* (c. 1128), like Cassiodorus' *On Sacred and Profane Learning*, allots an equal number of books to sacred and secular arts—in this case three to each; and like Cassiodorus, Hugh gives very little attention to quadrivium subjects. But in addition to the seven traditional *artes* handled by Cassiodorus, Hugh deals with physics and with the seven mechanical arts. Of interest to us is the chapter (III.2) listing the names of the founders, chief developers, and perfecters of the arts—a medieval writer's catalogue of the authorities he finds cited in his readings. In Hugh's list of nearly a hundred names, several are deities or mythical persons. According to Moses, the discoverer of music was Tubal, a descendant of Cain; according to some Greeks it was Pythagoras; according to others it was Mercury. The Egyptians invented geometry; Euclid, Eratosthenes, and Boethius were important in the transmission of the art. Cham, a son of Noah, was the inventor of astronomy. According to

Josephus, Abraham first taught the Egyptians astrology. King Ptolemy of Egypt was responsible for the revival of astronomy. The founder of medicine among the Greeks was Apollo. Medicine lay neglected for almost five hundred years, and then in the time of King Artaxerxes, Hippocrates reintroduced it.

It comes as a distinct surprise to find writers of the Roman period like Chalcidius and Macrobius attaining their greatest esteem as authorities on science at the very time when Latin translations of Greek and Arabic scientific works were being produced in full volume. These two Neoplatonists became the leading authorities on cosmography and geography in the twelfth century, and their popularity continued into the thirteenth and fourteenth centuries. Platonism, which afforded the Latin West its most vital contact with classical cosmography during the Middle Ages, did not reach its full flowering—chiefly at the school of Chartres—until the twelfth century. Latin Platonism had no direct contact with the master philosopher except through fifty-three chapters of Chalcidius' translation of the *Timaeus*. The most influential transmitters of medieval Platonism were Chalcidius, Macrobius, Augustine, Boethius, and Apuleius. Throughout this period Aristotle, because of Boethius' decision to translate his *Organon* first, was known almost solely as a dialectician. Not until translations of his scientific works from Arabic and Greek texts became available did Aristotle supplant Plato as the philosophers' inspiration on cosmography.

William of Conches was one of the three leading Platonists at Chartres, the other two being Thierry and Bernard Silvester. William was born in Normandy at the end of the eleventh century, may have been educated at Paris, and began teaching at Chartres around 1126. He undertook to reconcile Neoplatonic cosmography with the doctrines of the Church. He and his Chartrain colleagues did succeed in restoring Neoplatonism to good repute. William's greatest work was entitled *On the Philosophy of the Universe*. His *Dragmaticon philosophiae* was not a new work but presented the material of the larger work in the form of a dialogue between a philosopher and his pupil, the more readily to attract readers. William's works contain references to a commentary on the *Timaeus* which he prepared; and Victor Cousin discovered a manuscript of it in the Bibliothèque Nationale.

William's guides in astronomy were the Latin Neoplatonists Macrobius and Chalcidius. His doctrines on physics came from Isidore, Bede, and Hrabanus Maurus. At one point he cites Hipparchus, a mere name to him which he culled from Pliny, who also cited Hipparchus without handling his works. William's copying or close paraphrasing of Macrobius is exemplified in his remarks about the independent motions of the stars and the immense amount of time required for those motions to be detected, and in his discussion of the rival theories of the Egyptians and Chaldaeans regarding the correct location of the sun in the order of the planets. After pondering over Macrobius' enigmatic discussion, William comes to an interesting independent conclusion: since the sun, Mercury, and Venus have approximately the same periods, their orbits must be about the same size and, instead of being contained one within another, must intersect each other. He understands, from reading Pliny and Bede, that the planets' orbits are eccentric, but through some confusion he thinks the perigee of the sun's (apparent) annual course occurs in northern latitudes. He repeats Pliny's explanation of the planets' stations as resulting from the effect of the sun's rays arresting their courses and elevating or depressing the planets for a time. William follows the conventional view that the torrid and frigid zones are uninhabitable, but he rejects Macrobius' explanation that terrestrial zones have climates which correspond exactly to those of the celestial zones directly above them. Capella's egregious blunder that the known world and its antipodean region have common summers and winters crops up again here. Regarding the actual existence of antipodeans William adopts the cautious view, widely held by Church authorities, that the austral temperate belt is habitable but is not inhabited. He accepts Macrobius' notion of a great equatorial ocean encircling the globe, but in accounting for the tides he offers two explanations: the Cratean theory of the collision of ocean's streams at the poles, derived from Macrobius, and a theory resembling Adelard's of the oceans colliding against submerged mountains. The manuscript maps illustrating William's cosmographical works are Macrobian zone maps. Macrobius' crudely computed estimate of the sun's size as eight times the earth's size is acceptable to William.

At the middle of the twelfth century Hugues de Saint-Jean, a teacher at Chartres, and one Hugues Métel were corresponding about the priority of quality or quantity in relation to substance. Métel quotes Macrobius' statement to the effect that the first perfection of incorporeality is found in numbers. He goes on to say that, inspired by the example of

Scipio and Macrobius, he has become absorbed in mathematical studies —numbers, the dimensions of the earth, and the peculiar behavior of celestial bodies. He is not disturbed by the imputations of heresy that Manegold has attached to Macrobius' doctrines.

Thierry, another of the Platonist masters at Chartres, is an interesting transitional figure. One of his pupils, Bernard Silvester, dedicated his pagan-Christian cosmographical treatises to him; another pupil, Hermann the Dalmatian, dedicated to him his translation of an Arabic version of Ptolemy's *Planisphere*. Thierry stood waist-deep in the powerful crosscurrents of his day—Christian doctrine, Latin Platonism, and the new Greco-Arabic learning—and he tried to present a synthesis of all three. *On the Works of the Six Days* is his attempt to reconcile the scriptural account of the Creation with Platonic cosmography. He reproduces Macrobius' explanation that the universe, since it comprises all space and has no place outside where it can advance, must move in a rotary motion; he also reproduces Macrobius' circular reasoning to prove that the earth is at the center and bottom of the universe. Duhem feels that Thierry's comprehension of Aristotle's views is so much better than Macrobius' that he must have made use of a recent translation of an Aristotelian treatise on cosmography; but Duhem is in error in supposing that Macrobius obtained his ideas about the rotary motion of the universe from some Peripatetic source.[14] These views and their supporting arguments are stock material in the works of the Neoplatonic masters.

Thierry also compiled a large encyclopedia on the seven liberal arts, entitled *Heptateuchon*, which has never been published. The neglect of the work may perhaps be explained by the fact that it is an anthology of selected classical and early medieval treatises, many of them copied in toto. The collection follows the customary practice of devoting far greater attention to the trivium (434 sheets) than to the quadrivium (161 sheets). In the trivium are found all the then extant works of Priscian, two treatises of Donatus, various treatises of Aristotle and Cicero, the anonymous *Rhetorica ad Herennium*, and Boethius' treatise *On Hypothetical Syllogism* and his translation of Porphyry's *Introduction to the Categories of Aristotle*. The section on astronomy (77 sheets, or almost half of the entire book on the quadrivium) contains Hyginus' *Astronomicon* and Ptolemy's *Tables* and *Precepts*.

Thierry's pupil Bernard Silvester of Tours was strongly influenced by

Chalcidius and Macrobius and appears to have been more pagan than Christian in his cosmographical conceptions. *On the Universality of the World*, which he wrote during the papacy of Eugenius III (1145–53) and dedicated to Thierry, is divided into two books, on the megacosm and the microcosm. It purports to present Christian doctrines—in Neoplatonic terminology. The Neoplatonic Nous, which he declares to be the Providence of God, created the heavens and earth, and all the wonders of the earth. He gets his term for primary matter (*Sylva*) from Chalcidius. Nous also ordained that there be two temperate zones, cooled by the chill of the arctic zones and warmed by the heat of the torrid zone, where the sun keeps to a middle course—a Macrobian explanation of the climates. Macrobius was the inspiration of Bernard's other major work, a commentary on the first six books of the *Aeneid*. Like Macrobius, Bernard sees in Virgil not a master poet but a master philosopher, and the *Aeneid* is almost lost sight of as he uses it as pretext or occasion for sermons or indoctrinations. The commentary is full of allegorical fancies. The first six books of the *Aeneid* represent the six stages of man's life: for example, Aeneas at Carthage is a new-born soul, the burning of Troy in Book II represents the consuming passions of youth, and the visit to Hades in Book VI represents the departure to the afterlife. Dante drew some of his allegorical interpretations in the *Vita nuova* from this commentary.

About the year 1230 Bartholomew the Englishman compiled a bulky encyclopedia *On the Properties of Things*, in nineteen books. Its main interest lies in its zoological accounts, which are believed to have been the chief source of Shakespeare's natural history. The work consists almost entirely of reproductions and digests of information derived from earlier writers, whose names are scrupulously cited. In this pleasing respect Bartholomew resembles Vincent of Beauvais in his compilation of his vast encyclopedic works. Duhem's judgment about Bartholomew's book is a harsh one. He calls it a mediocre compilation displaying no ideas, unity, or critical faculties, a conglomeration of opinions on all knowable subjects—in short, a book with all the characteristics to assure it a great success.[15] About its vogue in the next few centuries there is no question; but other critics have praised Bartholomew for his personal observations, and his charming description of the domestic cat has elicited admiration from all sides. Such observations are juxtaposed with accounts of dog men and dragons from Pliny and of griffins from the medieval bestiaries.

Schedler has pointed out many instances of Bartholomew's appropriation of Macrobius' cosmographical doctrines, and observes that the thoughts of Macrobius are reported in an inexact manner, as if Bartholomew had been depending on his memory.[16] In a chapter (VIII.26) devoted to the planet Venus, Bartholomew reveals that he is interpreting Macrobius' ambiguous statements about the upper and lower courses of Venus and Mercury as being an exposition of the Heraclidean system. This mistake, we have noted, has persisted from the Middle Ages to the present time.

The heliocentric theory of Aristarchus of Samos was not noticed by any classical Latin writer, and there was no knowledge of it in the Latin West during the first Christian millennium. However, Duhem's investigations have revealed that heliocentricism did not die out completely during that time, for the Heraclidean theory of the heliocentric orbits of Venus and Mercury was transmitted throughout the Middle Ages to the time of Copernicus. A treatise on astronomy and astrology, *L'Introductoire d'astronomie*, written in 1270 in the prose of the Ile-de-France by the court astrologer of Baudoin de Courtenay, was examined in manuscript by Duhem.[17] The author pretends to be drawing his material from Ptolemy's *Almagest*, Aristotle, and Plato; his actual sources were Pliny, Macrobius, Capella, and William of Conches. Occasionally he credits Macrobius and Capella for his arguments. We find in this treatise a discussion of the Heraclidean geoheliocentric theory, related, as is usually the case, to the question of the correct order of the planets. Macrobius is cited on the contrary views of the Chaldaeans and Egyptians as to whether the sun should be placed immediately above the moon or above Mercury and Venus. The Heraclidean view is then introduced as a compromise attitude. The *Lucidator astrologiae* of Peter of Abano, written in 1310, is another work that presents the Heraclidean system.

We have emphasized the strongly antiquarian and pagan character of the scientific researches of the scholars at Chartres and the fact that they gave greater notoriety to Plato's *Timaeus* and its Latin Neoplatonic commentaries than those works had experienced in any earlier period of the Middle Ages. Chartrain antiquarianism was not another resurgence of the stale brand of antiquarianism that tainted almost all of early medieval literature. It was rather a manifestation of the sweeping purviews of scholars who would more properly be regarded as innovators than as antiquarians. Chartres soon became a center of Arabic studies. It has been argued that some of Avicenna's commentaries were known even to Thierry and William of Conches. Chartres scholars became intimately acquainted with the work of the translators of Greek and Arabic texts in Spain and southern Italy, and Chartres was the first school in northern Europe to introduce studies of Ptolemy's works and the newly translated treatises of Aristotle.

As the scientific works of Ptolemy and Aristotle came to be studied intensively and with understanding, enthusiasm for Latin handbook science began to wane; but the transition was a gradual one. Olaf Pederson points out that of the 116 translations which Sarton attributes to Adelard of Bath, Gundisalinus and his school, and Gerard of Cremona and his pupils, 84 have to do with the quadrivium.[18] Some branches of the old Latin science persisted with surprising hardiness: Pythagoreanism, which had a vogue into modern times; medicine and veterinary medicine, which for practical reasons continued to be subjects of study throughout the Middle Ages; meteorology; and geography, which, from ancient Greek times, was always one of the most backward of the sciences. Ptolemy's *Geography* was not translated until 1410, long after most of his less important treatises had been translated, some of them several times, from both Greek and Arabic. In the meantime the inept geographical accounts of Pliny, Solinus, Capella, and Isidore served to give medieval readers a meager and distorted impression of the state of geographical knowledge among the ancient Greeks. Most Latin geographical writers before Ptolemy's work became available still thought of the known world as surrounded by ocean and of the torrid zone as uninhabited. The regressive tendencies of geographers continued to be manifested in the Renaissance and into modern times. Not long after Ptolemy became the new authority on geography, many of his basic conceptions were disproved by the voyages of Columbus and Magellan. Even so, readers preferred atlases in the Ptolemaic style, and cartographers had to accommodate their maps to the new tastes with a curious blend of Ptolemaic authority and the newly acquired information of explorers. The atlases of Mercator and Ortelius discarded most of the Ptolemaic errors, but some persisted in maps in the eighteenth and nineteenth centuries.

In examining representative texts of medieval Latin science, from

Boethius to the school of Chartres, we are disheartened to observe the same appalling traits and scholarly practices being exhibited by medieval writers that prevailed among classical and postclassical Latin writers on science. For well over a thousand years Western intellectuals lacked the motivation to come to grips with the Greek scientific disciplines and were content with creating an impression of scientific learning and acumen. During the last century of the Roman Republic and the period of the western empire abundant means of travel and communication existed, under Roman provincial administration, between East and West. Latin writers could have informed themselves of the results of the researches conducted at Alexandria and other Greek centers of science. Medieval writers also had opportunities to learn about Greek science through the channels of the Church. Yet we rarely find a Latin writer in either age who indicates an inkling of understanding of the vast difference between the levels of Greek and Roman science.

We might be inclined to suppose that in certain periods medieval scientists had closer contacts with ancient Greek science than the Romans had. The Middle Ages had Boethius' adaptation of Ptolemy's *Harmonics* and his translation of Nicomachus' *Introduction to Arithmetic*, as well as fragments of his handbook on Euclidean geometry; and there were traces of ancient Greek astronomy in the computus literature. Remarks in the writings of Cassiodorus and Bede suggest that Greek scientific treatises were available in libraries and that scholars were able to comprehend them; but if such opportunities did exist, virtually no tangible results were forthcoming. To surmise, on the basis of extant texts, that the most gifted of the early medieval writers were more competent to understand Greek science than the classical Latin writers would be unwarranted. Allowance must be made for the fact that medieval texts are more abundantly preserved than Roman texts. The scientific knowledge of Varro and Boethius cannot be compared because of the loss of Varro's writings on the quadrivium. But fate has also been unkind enough to preserve some of the most benighted specimens of medieval scientific writing. It is evident that in the darkest periods a scholar's reputation for being the most learned man of his age was based upon the fact that he could read and write Latin and had the energy and enterprise to copy or paraphrase the compendia that he found ready at hand.

XV

CONCLUSION

For those who are interested in tracing developments in theoretical science, the Roman period as we have seen is barren; but it is as serious a mistake for a historian of science to neglect the Latin writers on scientific subjects as for a student of European history to neglect secular intellectual developments when studying the Dark Ages. A highly regarded history of science from St. Augustine to Galileo (A.D. 400–1650), which has recently been revised in two volumes,[1] may be taken as typical of the tendency to pass lightly over the period of Roman influence: it devotes only the first chapter to the Latin science of the West before the Greco-Arabic Renaissance in the twelfth century. Nor should our admiration for the Romans induce us to conceal their ineptitudes in science by shifting our attention to their technology or by highlighting scientific discoveries that happen to have been first recorded by Roman writers though actually made by the Greeks. To do so is to create a false impression that Roman science was passably good.

Derivative and stereotyped as the Latin compilations may be, we must examine them in order to understand a lengthy period of intellectual history. If we wish to learn why science and philosophy were so retarded in Western Europe in the early Middle Ages, we must go back to ancient times for an explanation; most of the manifestations of low-level scientific and philosophical thinking that we associate with the Dark Ages appear among the Romans; and before their day we find clear indication of superficial and pretentious scholarly habits developing among the Hellenistic Greeks. How far back this shoddiness goes can only be guessed, but the evidence of it grows clearer as time passes. The symptoms evident in the Hellenistic Age get progressively worse in the Latin West, with occasional rallies, as in the cases of Boethius and Bede, until the Greco-Arabic influences begin to be felt.

Scientific knowledge in Roman times never rose above the lowest level of Greek popular science. Scientific matters were left mainly to compilers and encyclopedists. The Latin writers who enjoyed reputations as scientific authorities were not the most gifted prose stylists, like Cicero and Tacitus, but such erudites and polymaths as Varro, Nigidius Figulus, and Pliny. The point may be raised that polymathy has frequently been associated with eminent scientists, in every age until the present age of specialization, and that the ancient Greeks were no exception. Some of their leading scientists—Aristotle, Theophrastus, Eratosthenes, Ptolemy, and Galen—had overpowering reputations as polymaths as well. But there should be no confusion between a Greek polymath who made vital contributions to scientific progress and a Roman encyclopedist who was a mere compiler, skimming the surface of the scientific disciplines.

There was probably a closer kinship between Greek and Roman writers on the popular-science level than between the Greek popular writers and their fellow countrymen who were the true experts. Examples that illustrate the difference between the high and low levels of Greek scientific writing are easy to find. Take Euclid's *Elements*, one of the finest and most successful textbooks ever produced. It contains no pretense, no flowery introduction, no protreptics to the virtuous life. The opening sentences are familiar: "A point is that which has no part; a line is length without breadth; the extremities of a line are points." Or take the works of Archimedes, one of the pre-eminent mathematicians of all time, which are usually written in the form of letters to colleagues or pupils and open with a salutation; "Archimedes to ———, greeting." The gist of his introductory remarks is likely to be: On a former occasion I sent you several propositions and proofs, showing that . . . Since that time I have been troubled about certain problems. Recently certain theorems, not hitherto demonstrated, have occurred to me and I send you the proofs herewith. They are as follows . . .

Works like these command the respect of scientists today. They are things of beauty, both in their logic and in their systematic form of presentation, and stand like beacons of approach to a subject. The thoughts are brilliant, but the creator is a humble man, interested only in recording his discoveries accurately. Compare these openings with the introductory chapters of the book which the ancients regarded as the classic work in its field, the *Introduction to Arithmetic* of the popularizer Nicomachus of Gerasa, which devotes six chapters to the definition and goals of philosophy and the importance of mathematical instruction for an understanding of the nature of the universe. Or contrast the opening sentences of Ptolemy's *Geography*, which define the term geography and explain the difference between it and chorography, with the first chapter of that other extant ancient classic on geography by the non-mathematician Strabo. Strabo claims geography as a province of the philosopher—Homer to him was a philosopher. The geographer, he says, is naturally a polymath. Geography explores everything in the heavens and on earth, is useful to generals and politicians, as well as to philosophers, and ultimately relates to the happy life. The prefatory remarks of Strabo and Nicomachus, instead of reflecting a scientific attitude, anticipate the medieval attitude that regarded any sort of "learned" (L. *doctus*) man—poet, mining engineer, teacher of grammar or geometry—as a philosopher.

In this volume we have had occasion to emphasize tedious and fraudulent scholarship—the copying of texts and the indiscriminate appropriation of readily digestible materials without acknowledgment or with misrepresentation of sources. These disclosures should not come as a disappointment to admirers of classical and medieval culture. We must bear in mind that we have been dealing here mostly with transmitted and received doctrines gathered by scholars who were seldom capable of original ideas or observations. Our account of Roman science, unfortunately, has not been alleviated by reflections upon the exceptional accomplishments of classical and medieval technicians and engineers, artists and craftsmen, in whose productions we would find much to admire; nor do the lay compilations that we have encountered exhibit genuine literary merit. In classical Greek and Roman society surpassing genius and inspiration were lavished upon works of beauty, in the arts and crafts as well as in literature. But lay handbook science was virtually devoid of literary merit, originality, and direct contact with natural phenomena; instead the popular handbooks resembled school textbooks, which, in the ancient and medieval worlds, were notoriously dull and full of commonplaces.[2]

Primary and secondary schooling had become fairly standardized as early as the beginning of the Hellenistic Age.[3] Embryonic philosophers, scientists, poets—and compilers—were alike subjected to much the same type of curriculum. But at the conclusion of their formal studies those with superior artistic or intellectual endowments rose above the level of the schools, whereas men with scholarly bent but little talent or imagination took up, if they must, the impecunious profession of teaching, or continued in their studies and writing to exercise their predilections for the assimilable materials of textbooks and commentaries.

For most students in the ancient and medieval worlds formal schooling must have been a humdrum experience, with much attention directed to absorbing principles of rhetoric and to dissecting classical poetry and prose. Among the esteemed goals of a classical education were adeptness in literary criticism and an impressive rhetorical style; not proficiency in the sciences. Learning was mainly by rote, sternly applied. When the teachers were pedants, as was generally true of the professionals, the works of classical authors were scrutinized for rhetorical tags, evidences of erudition, and moral lessons. Uninspired schoolmasters, insensitive to poetic beauty, regarded Homer, Hesiod, or Virgil merely as men of prophetic vision and uncanny scholarship. The fanciful attribution of erudition to the poets provided such teachers and commentators with an opportunity to display their own jejune erudition.

Pedantry was not confined to the schools in the Hellenistic and Roman worlds. The intellectual world into which the more brilliant youth were being introduced cherished ostentatious learning, and polymaths were the elite of that society. The desire to impress one's fellows by embracing all fields of knowledge and to be able to explicate difficult literary and philosophical works was so keen that there was a steady demand for introductory manuals in the academic subjects. Men of talent, as well as pedants, responded on occasion to this general craving for erudition by writing textbooks. A creative scientist might produce a manual that could be comprehended by laymen—Eratosthenes' lost *Platonicus* was probably such a book—or systematize a scientific discipline in the form of a handbook, as Euclid did in his *Elements* and Ptolemy in his *Almagest.*

The poets, too, were responsive to popular desires to acquire learning, and aware that versifying would lend attraction to dry contents of handbooks. Many poets turned their attention to didactic subjects, in spite of the fact that they could not comprehend scientific material written above the layman's level. Virgil derived his astronomical lore from his poet-predecessor Aratus, not from Hipparchus and Eratosthenes.[4] While some didactic poets, like Lucretius and Virgil, were able to produce works of rare beauty and inspiration, others, poetasters such as Nicander, ground out poetic versions of handbooks that merely catered to a taste for verse.

The compilers, too, had regard for the tastes and habits of the reading public. They appreciated that they could gain a wider audience by identifying their writings with famous works of literature. Thus they produced volumes that were ostensibly commentaries on, or introductions to, favorite classics; but in some cases the outward form was a maladroit disguise for a conventional manual; in others passages of text served as excuse for embarking upon excursuses that were actually segments of handbooks.

To both Greek and Roman popular writers, science had an aura of uncanniness. It seemed to them related to matters of revealed knowledge, the occult, the exotic, the arts of the astrologer and soothsayer. Nature appeared to them, as to initiate worshipers of the Eleusinian and Orphic mysteries, to be jealously guarding her secrets. These might be disclosed as readily by poets and philologists as by speculative philosophers and men who had actually investigated natural phenomena. Opinions ascribed to early sources came to carry greater weight than those of presumably less gifted contemporary men, and were enhanced with a mystical quality if the traditions were oral, not written. This attitude had much to do with the widespread practice of attributing discoveries of say the Hellenistic Age to remote personages like Asclepius, Homer, or Pythagoras.

If we are inclined to poke fun at the ludicrousness of early medieval Latin science, we had better remember Pliny, who claims to have examined about two thousand works in preparation for writing his *Natural History.* Among the 473 primary authorities that he vaunts in his bibliography we find not only Greek but Egyptian, Babylonian, Persian,

Etruscan, and Carthaginian names. He lists as his Greek sources for the astronomical portion of his work Hipparchus, Sosigenes, Posidonius, Anaximander, Eudoxus, Euclid, Dicearchus, Archimedes, and of course Eratosthenes—yet he had only the faintest inkling of the achievements of these mathematicians. Chalcidius in the fourth century pretended to be using in his commentary on the *Timaeus* forty-eight pre-Christian authorities—such as Zoroaster, Orpheus, and Pythagoras—and never mentioned Theon of Smyrna, whose manual he was freely translating. By the Middle Ages Adam, Abraham, and Prometheus were being cited as authorities on astronomy. Naming sources that no reader had encountered before made a writer appear recondite; and the reputation of antiquarian was assurance that he was in possession of true wisdom.

The contacts with the masters of Greek philosophy and science became increasingly remote as one compiler after another drew upon the work of his predecessors to produce volumes even more easy to assimilate. In generation after generation compilers bandied the names of Archimedes, Hipparchus, Ptolemy, Plato, and Aristotle without knowing who came first or whether they were all contemporaries. Since being able to cite the names of the great minds of the ancient world gave an aura of authority, the names, like trademarks of the profession, are usually spelled correctly. But still "Eratosthenes" can get thoroughly mangled by a medieval scholar who pretends to be criticizing his procedures in measuring the earth's circumference. In the Migne text John Scot Erigena requires three columns of close print for his pretentious explanation of how Eratosthenes measured the globe—and all the while he is himself assuming that a circumference is equal to twice the diameter. Erigena's blunder probably resulted from misinterpreting a statement made by Martianus Capella, but Capella in turn did not have the remotest idea what Eratosthenes was doing. Nor did any other ancient Roman so far as we know understand the simple geometrical demonstration involved in Eratosthenes' method.

Textbooks on scientific subjects might be conceived as little more than repositories of discrete, compartmentalized revelations. A common source of information was the various collections of the opinions of philosophers, such as the main doxographic collection, which originated in the large compilation prepared by Theophrastus, and the so-called biographical doxographies. That the doctrines of the doxographers were transmitted by laymen as separate bits of dogma may account in no small measure for the strange incongruity of opinions, some startlingly perceptive and others childishly naive, that have been attributed in the doxographic collection to the same early philosopher. It is not a little discomforting to realize in how many cases what have come down to us as the opinions of eminent philosophers and gifted writers were transmitted and preserved by ill-informed, careless, and untrustworthy compilers.

The very discreteness of the data and the uncritical mentality of the men who gathered them ought to admonish us to caution. A modern scholar who handles any quotation in an ancient compilation as if it were an authentic document from the period of the compiler, or who chooses any isolated datum (often inconsistent with other data in the same work) to use as the crux of an argument or the basis for a theory, may be showing no more discernment than his predecessors the compilers. Diligence, both ancient and modern, has nearly always been rewarded by the discovery of some nuggets that have served to support a preconceived theory.

In falsifying their sources the ancient and medieval encyclopedists succeeded in imposing not only upon readers of their own time but upon subsequent generations of scholars up to the present. It may amuse us to discover that a writer like John of Salisbury (c. 1115–80) derived the doctrines about suicide in his *Policraticus* from Cicero and Macrobius but attributed them to Pythagoras and Plotinus, without realizing that Macrobius, who also attributed them to Plotinus, had in fact obtained them from Porphyry.[5] It is less amusing to find that we have ourselves been deluded by such subterfuges. Not until the present century have scholars generally appreciated that the citations made by Latin encyclopedists, early and late, were usually fallacious. Perceptive investigators like Ludwig Traube and Pierre Courcelle have pointed out that the frequent use of Greek words and phrases by medieval Latin writers is not to be taken as evidence that they were all able to read Greek; often the same words and phrases turn up in the works of several writers; mutilated and mispelled forms recur; and it becomes plain that successive borrowers were perpetuating a sham. Medieval historians have been more generally aware of this fact than historians of science. Ptolemy and Galen are still categorically referred to as the great authorities of medieval science, although their influence in the West remained negligible until their works had been translated from Arabic and Greek in the later Middle Ages; citations of the works of Aristotle, Archimedes, Eratosthenes, and Hipparchus made by Latin writers before the Greco-Arabic revival continue to be treated as if they were trustworthy.

Even direct quotations are not to be taken as conclusive evidence that a compiler had handled a cited work. Greek compilers were fond of seeming to be quoting Plato, but their quotations usually prove to be inaccurate, owing to careless transcriptions by generations of Platonists. These extracts became commonplace tags, cited repeatedly by successive authors, and in the same sequence. Direct quotations from the *Enneads* of Plotinus—the bible of Neoplatonism but a notoriously difficult work to understand—are not to be considered as proof that his fervent disciples had read his work. They too appear from their frequent occurrence to have been favorite sayings, and a large number of them come from opening passages of a book or chapter. The Neoplatonic doctrines that were circulated in the Latin West were not derived from Plotinus but came, usually indirectly, from Porphyry or some other readily intelligible disciple. In subjects like theoretical science and philosophy, in which the Romans freely admitted the superiority of the Greeks, Latin writers naturally wished to gain authority for their manuals by appearing to be handling Greek works. But the actual evidence of firsthand knowledge of Greek science in the Latin West during the early Middle Ages is very slight.

Occasionally it has been suggested that we should condone the pilferings and subterfuges of the ancients, in their handling of source materials, as the accepted practice and conventional morality of the times. That the practices were widespread is true, but there is also no denying that the perpetrators had some sense of the morality and guilt involved. Pliny's laudable sentiments in the Preface of his *Natural History* about the proper attitude of a gentleman regarding the acknowledgment of his sources were referred to earlier and deserve fuller quotation here:

> The fact that I have prefaced these volumes with the names of my authorities will be proof to you of my pride and spirit. For I consider it a matter of courtesy, honor, and modesty to acknowledge the names of one's sources and not to follow the example of most of the authors whom I have handled. I want you to know that when I was comparing my authorities, I discovered that the most recent and highly reputed authors have copied word for word from the texts of early authorities and have withheld their names. . . . Surely it is the mark of a mean spirit and a wretched nature to prefer to be caught in a theft rather

than to repay a loan—especially considering the fact that principal grows out of interest.

There are numerous cases in extant works of deliberate suppression of the name of an author whose book was being appropriated in whole or major part. On the other hand, scrupulous acknowledgments of sources were made by Cassiodorus and the Venerable Bede in some of their later works; and in the twelfth and thirteenth centuries Lambert of St. Omer, Bartholomew the Englishman, and Vincent of Beauvais displayed gentlemanly scruples about crediting their sources.

Their pretense of laborious and recondite scholarship cannot conceal the fact that most of the Latin encyclopedists were mentally rather indolent fellows, incapable of the sustained concentration required to master any subject. A sampling of opinions culled from a popular collection or from an abstract of an earlier work often satisfied their curiosity on a subject. A handful of basic definitions, axioms and propositions, without proof—extracted in an earlier century from some abridgment of Euclid's *Elements*—sufficed for a treatment of geometry; and a few Pythagorean tenets about planetary distances being determined by a musical scale took care of one of the more difficult topics in a discourse on astronomy. The Latin science handbooks presented stock materials—threadbare doctrines, specific data, commonplace examples, and quotations from classical poets. From the significant proportion of excerpts that turn out to be opening phrases or lines, it seems likely that the scholar who culled them merely glanced at the beginning of a scroll without unrolling it or scanned headings without reading the text. It was easy to appropriate an existing treatise or two and pass off a paraphrased version as one's own treatment of a subject or discipline. Topics then were usually taken up in their original order; pages of text might be lifted verbatim or translated literally; and quotations from otherwise unknown poets were taken over with the borrowed passages. By citing several authorities, sometimes putting them in opposition to each other, or by ridiculing the garbled views of one or two of them, the compiler hoped to conceal the derivative character of his book.

The more gifted Roman intellectuals of the classical period failed to appreciate the systematic nature of scientific disciplines and showed no disposition to master any of the Greek sciences. They dipped into theoretical treatises only on rare occasions and generally left such matters to hacks and polymaths, who had even less understanding of the rigor and logic of scientific investigation. Because of their lack of interest in scientific thought the Romans must be held responsible for the deteriorated state of knowledge in Western Europe during the first millennium of the Christian Era. While in the eastern empire Byzantine and Alexandrian scholars continued to study the original classics of Greek science, the compilers of the Latin West rummaged no further than the readily digestible compendia of their most recent predecessors. As a consequence, the works of the postclassical compilers have often survived while the classical Latin works which were their vernacular antecedents have perished.

We must not suppose that the Romans were incapable of gaining some understanding of Greek science. There were, to be sure, Latin scholars of the late republic and early empire who could have translated most of the Greek scientific classics—if not with accuracy and real comprehension of the theoretical portions, at least well enough to give perceptive readers some insights into the logical nature of scientific thought. We have seen how few were the instances of Latin scholars undertaking the translation of significant Greek treatises. Apuleius translated Nicomachus' *Introduction to Arithmetic* in the second century; during the brief revival of Greek studies that was instituted by Theodoric and continued after his death, Boethius translated Euclidean, Nicomachean, and Ptolemaic treatises; and Cassiodorus set his monks to the task of translating secular as well as religious works for the library collection at Vivarium. Cassiodorus refers with pride, in *On Sacred and Profane Learning*, to the translations of several of the Greek classics in mathematics, astronomy, and medicine that had been made available under this program; but Cassiodorus' glowing reflections on the state of science and learning at his monastery must be regarded with reservations in the light of his evident exuberance when writing about the scientific attainments of his colleagues. In the East, meanwhile, the Greek classics were being zealously preserved, studied and recopied in each generation; and commented upon, translated, and abstracted to make them more accessible to a wide reading public. The Arabs came into possession of these works intact, by translating them directly from Greek or by using intermediate Syriac or Hebrew versions. Throughout the expansive Arabic world, including Sicily and Spain, the masterpieces of Greek science could be read in complete and unadulterated form.

After the lapse of centuries the gifted minds of Western Europe had a second opportunity to acquaint themselves with the Greek classics, preserved substantially in their original form. Largely stimulated by their contacts with Arabic civilization, awakened medieval scholars like Adelard of Bath explored the "new science." By the twelfth century, when the excitement of discovery reached its height, keen-minded individuals had a genuine appreciation of the fundamental differences between a treatise *On Optics* by Ptolemy and Plinian anecdotes about mirrors and magnifying glasses. In their passionate desire to embrace Greco-Arabic science some scholars, such as Gerard of Cremona and Michael Scot, traveled to Spain or Sicily to learn Arabic and to ferret out manuscripts; others worked with interpreters or with vernacular versions of Arabic versions of a Greek original. It was far more difficult to master Greek science in this roundabout way than it would have been for the ancient Romans, who, in the classical period, had been taught to read Greek from early schooldays, who sometimes put the finishing touches upon their formal education by studying in Greece, and could go to the original works.

As we reflect upon the Romans' attitude toward scientific knowledge and their handling of it, we glimpse the significance of their failure to transmit original Greek science to the Western world. It assumes a tragic aspect and may rightly be regarded as one of the greatest deficiencies in their civilization. Intellectual life in Western Europe would have been vastly different during the early Middle Ages if the Romans had been interested in mastering, instead of dabbling in, Greek theoretical studies.

As far as the West is concerned, the later Middle Ages have more in common with the Renaissance than with the early Middle Ages. It would seem to be more sensible to refer to Latin traditions of science and philosophy as medieval, and to Greco-Arabic traditions as Renaissance. Existing congenially side by side for several centuries, although they were different in spirit and substance, they baffle our attempts to arrive at a satisfactory chronological demarcation between the Middle Ages and the Renaissance.

REFERENCES AND BIBLIOGRAPHY

The references listed below are intended to provide the reader with some of the primary evidence and some views of authoritative scholars bearing upon statements made in this book. Many of them will also serve as a guide to further reading on the subject. The bibliographies are selected lists of reference works, books, and articles that have been consulted in the preparation of this volume. The following abbreviations have been used:

P–W: Pauly–Wissowa, *Real-Encyclopädie der classischen Alter-tumswissenschaft*

S–H: Schanz–Hosius, *Geschichte der römischen Literatur*

S–S: Schmid–Stählin, *Geschichte der griechischen Literatur*

T–S: Teuffel–Schwabe, *History of Roman Literature* (Warr translation)

The bibliographies and discussion of pertinent matters contained in George Sarton's *Introduction to the History of Science*, M. R. Cohen's and I. E. Drabkin's *A Source Book in Greek Science*, and *The Oxford Classical Dictionary* have been consulted throughout.

I

INTRODUCTION

1   Edward Gibbon, *The History of the Decline and Fall of the Roman Empire*, chap. 13; ed. by J. B. Bury (10th ed. London, 1930), 1:365.

2   On the superiority of Roman engineers, see R. J. Forbes, *Man the Maker; a History of Technology and Engineering* (London and New York, 1958), 69–71; on the small amount of mathematics required in practical applications, see O. Neugebauer, *The Exact Sciences in Antiquity* (Providence, 1957), 71–72, 80; J. L. Heiberg, *Mathematics and Physical Science in Classical Antiquity* (London, 1922), 81; on the small number of scholars engaged in scientific research, see O. Neugebauer, "Exact Science in Antiquity," *University of Pennsylvania Bicentennial Conference: Studies in Civilization* (Philadelphia, 1941), 25; reprinted in *Toward Modern Science*, ed. by R. M. Palter (New York, 1961), 1:17; on the scientific acumen of the Greeks being largely devoted to the development of the arts of warfare, see *A History of Technology*, ed. by Charles Singer and others (Oxford, 1956), 2:699; on the indebtedness of Roman applied science to the Greeks, see ibid., 116–17, 635–36; on the relationship between science and technology, see M. R. Cohen and I. E. Drabkin, *A Source Book in Greek Science* (Cambridge, Mass., 1958), 182–256, 314–51; R. J. Forbes, *Studies in Ancient Technology* (Leiden, 1955), 2:97–98.

3   On Roman dependence upon Greek technicians, see Heiberg, *Mathematics and Physical Science*, 81; on the difficulties in tracing the development of instruments and the transmission of technological knowledge, see G. Sarton, *A History of Science; Hellenistic Science and Culture in the Last Three Centuries B.C.* (Cambridge, Mass., 1959), 117–18, 377; *Ancient Science and Modern Civilization* (Lincoln, University of Nebraska Press, 1954), 44; on the technical heritage of the Greco-Roman

world being transmitted to the medieval world, see Forbes, *Man the Maker*, 107.

4   See the extant fragment (No. 1), trans. by Kathleen Freeman, *Ancilla to the Pre-Socratic Philosophers* (Cambridge, Mass., 1957), 78.

5   See H. Einbinder, "A Straight Look at the Encyclopaedia Britannica," *Columbia University Forum*, 3 (1960): 20–25.

BIBLIOGRAPHY

Georgius Agricola, *De re metallica libri XII*; English trans. by Herbert C. and Lou H. Hoover. New York, 1950.

K. H. Dannenfeldt, "Egypt in the Renaissance," *Studies in the Renaissance*, 6 (1959): 7–27.

M. Fuhrmann, *Das systematische Lehrbuch; ein Beitrag zur Geschichte der Wissenschaften in der Antike*. Göttingen, 1960.

Orth, "Bergbau," P–W, *Supplementband*, 4 (1924): 108–55.

A. P. Usher, *A History of Mechanical Inventions*. Cambridge, Mass., 1954.

L. White, "Technology and Invention in the Middle Ages," *Speculum*, 15 (1940): 141–59.

II

CLASSICAL GREEK ORIGINS

1   On the high regard in which Pythagoras' doctrines were held by Renaissance writers, see S. K. Heninger, Jr., "Some Renaissance Versions of the Pythagorean Tetrad," *Studies in the Renaissance*, 8 (1961): 7–33.

2   See H. I. Marrou, *A History of Education in Antiquity* (New York, 1956), 47.

3   See the extant fragment (No. 40), trans. by K. Freeman, *Ancilla to the Pre-Socratic Philosophers*, 27.

4   On the reasoning that if numbers are the key to harmony they may be the key to the universe, see J. Burnet, *Early Greek Philosophy* (4th ed. London, 1930), 107; S. Sambursky, *The Physical World of the Greeks* (London, 1956), 38–40; W. Jaeger, *Paideia; the Ideals of Greek Culture* (New York, 1945), 1:164–65; T. L. Heath, *A History of Greek Mathematics* (Oxford, 1921), 1:68–69; on the possible connection between the seven notes of the heptachord and the seven visible planets, see G. Sarton, *A History of Science; Ancient Science through the Golden Age of Greece* (Cambridge, Mass., 1952), 214 [n].

5   See Nicomachus of Gerasa, *Introduction to Arithmetic*; English trans. by M. L. D'Ooge, with studies in Greek arithmetic by F. E. Robbins and L. C. Karpinski (Ann Arbor, Mich., 1938), 18–19; Burnet, *Early Greek Philosophy*, 106; Heath, *A History of Greek Mathematics*, 1:99.

6   Nicolaus Copernicus Thorunensis, *De revolutionibus orbium caelestium libri VI* (Torun, 1873), 34. On Copernicus' deletion of the passage about Aristarchus, see T. W. Africa, "Copernicus' Relation to Aristarchus and Pythagoras," *Isis*, 52 (1961): 406–7.

7   They are so regarded by Jaeger, *Paideia*, 1:316–17, and Marrou, *History of Education*, 54–56.

8   This is the opinion of Jaeger, 294, and Marrou, 48–49.

9   *Platonism Ancient and Modern* (Berkeley, University of California Press, 1938), 105.

10   *A Commentary on Plato's Timaeus* (Oxford, 1928), 2.

11   C. L. Ideler, "Über Eudoxus," *Abhandlungen der Berliner Akademie, historisch-philosophische Classe* (1828), 189–212; (1830), 49–88; G. Schiaparelli, "Le sfere omocentriche di Eudosso, di Calippo e di Aristotele," *Pubblicazione del R. Osservatorio di Brera, Milano* No. 9.

12   The Meno fragments have been edited and translated by W. H. S. Jones, *The Medical Writings of Anonymus Londinensis.* Cambridge, 1947.

BIBLIOGRAPHY

G. J. Allman, *Greek Geometry from Thales to Euclid* (Dublin and London, 1889), for Pythagoras, Archytus, Eudoxus, and Theaetetus.

Aristotle, *De caelo*; trans. by J. L. Stocks, in *The Works of Aristotle*; trans. into English under the editorship of W. D. Ross (Oxford, 1922), Vol. 2.

H. Cherniss, "Plato as Mathematician," *Review of Metaphysics*, 4 (1951): 395–425.

F. M. Cornford, *Plato's Cosmology; the Timaeus of Plato*; trans. with a running commentary. London, 1937.

——— *Plato's Theory of Knowledge; the Theaetetus and the Sophist of Plato*; trans. with a running commentary. London, 1935.

J. L. E. Dreyer, *History of the Planetary Systems from Thales to Kepler* (Cambridge, England, 1906; reprinted New York, 1953), for Eudoxus.

P. Duhem, *Le système du monde* (Paris, 1913–). Vol. 1 for Philolaus, Hicetas, Ecphantus, Plato, and Aristotle.

T. L. Heath, *Aristarchus of Samos* (Oxford, 1913), for Plato and Eudoxus.

Hipparchus, *In Arati et Eudoxi Phaenomena commentariorum libri III*, ed. by K. Manitius. Leipzig, 1894.

"Hippias," S–S, Pt. 1, 3 (1940): 49–57.

W. Jaeger, *Paideia*. Vol. 1 for the Sophists, Plato, and Isocrates.

J. S. Morrison, "Pythagoras of Samos," *Classical Quarterly*, 50 (1950): 135–56.

J. A. Philip, "The Biographical Tradition—Pythagoras," *Transactions of the American Philological Association*, 90 (1959): 185–94.

"Pythagoras; ältere Pythagoreer," S–S, Pt. 1, 1 (1929): 732–41.

E. Sachs, *Die fünf platonischen Körper; zur Geschichte der Mathematik und der Elementenlehre Platons und der Pythagoreer.* Berlin, 1917.

B. L. van der Waerden, *Science Awakening*; trans. by A. Dresden, with additions of the author. Groningen, 1954.

III

HELLENISTIC HANDBOOK TRADITIONS

1   See W. W. Tarn, *Hellenistic Civilization* (3d ed., rev. by the author and G. T. Griffith, London, 1952), 167–70, 189–202, 268–70, 295; G. Sarton, *History of Science in the Last Three Centuries B.C.*, chaps. 2, 10, et passim; W. H. S. Jones, "Hellenistic Science and Mathematics," *The Cambridge Ancient History* Vol. 7 (1928), chap. 9.

2   A good summary of Pytheas' achievements is to be found in *The Geographical Fragments of Hipparchus*; ed. with an intro. and commentary by D. R. Dicks (London, 1960), 179–82.

3   See "The Development of Greek Anatomy," *Bulletin of the Institute of the History of Medicine, Johns Hopkins University*, 3 (1935): 235–48.

4   Timon of Phlius, quoted by Athenaeus, *Deipnosophistae* I.22d.

5   *Doxographi Graeci* (Berlin, 1879); subsequently revised in several editions.

6   *History of Greek Mathematics*, 1:118. B. L. van der Waerden, *Science Awakening*, 91, is of the same opinion.

7   *Ancient Science and Modern Civilization*, 24.

8   Dicks criticizes the view of Rehm (which Heiberg also had expressed in his *Mathematics and Physical Science*, 71) that Hipparchus' commentary can be lightly regarded as a youthful work. See *The Geographical Fragments of Hipparchus*, ed. by Dicks, 11.

9   George Sarton was puzzled about Eratosthenes' contemporary reputation as a philologist when, according to Sarton, he should have been known as a scientist. See his *History of Science in the Last Three Centuries B.C.*, 110.

10   On the problems involved in attempting to find a correct value for the stade used by Eratosthenes, see *The Geographical Fragments of Hipparchus*, ed. by Dicks, 42–46.

11   Menecles of Barca and Andron of Alexandria, quoted by Athenaeus, *Deipnosophistae* IV.184b–d.

BIBLIOGRAPHY

Aratus, *Phaenomena*; ed. by E. Maass. 2d ed. Berlin, 1955.

"Aratus," S–S, Pt. 2, 1 (1920): 163–67.

Autolykos, *Rotierende Kugel und Aufgang und Untergang der Gestirne*, and Theodosius of Tripolis, *Sphaerik*; trans. from the Greek, with notes, by A. Czwalina. Leipzig, 1931.

E. H. Berger, *Die geographischen Fragmente des Eratosthenes*. Leipzig, 1880.

G. E. Broche, *Pythéas le Massaliote, découvreur de l'Extrême Occident et du nord de l'Europe*. Paris, 1935.

E. H. Bunbury, *A History of Ancient Geography*. 2d ed. London, 1833; reprinted New York, 1959. Vol. 1 for Pytheas and Eratosthenes.

Callimachus and Lycophron; Aratus; English trans. by A. W. and G. R. Mair. London, 1921.

Commentariorum in Aratum reliquiae; ed. by E. Maass. Berlin, 1898.

A. Couat, *Alexandrian Poetry under the First Three Ptolemies* (London and New York, 1931), for Aratus.

The Geographical Fragments of Hipparchus, ed. by Dicks, for Pytheas, Eratosthenes, and Hipparchus.

J. L. E. Dreyer, *History of the Planetary Systems*, for Heraclides, Aristarchus, and Eratosthenes.

P. Duhem, *Le système du monde*. Vol. 1 (1913) for Heraclides and Aristarchus.

The Thirteen Books of Euclid's Elements; trans., with an intro. and commentary, by T. L. Heath. 2d ed. Cambridge, 1926; reprinted New York, 1956. Vol. 1 for a discussion of Euclid's works.

F. Gisinger, "Geographie (Eratosthenes)," P–W, *Supplementband*, 4 (1924): 604–14.

———— "Geographie (Krates)," ibid., 614–17.

———— "Oikumene (Krates)," P–W, 17 (1937): 2143–45.

———— "Perioikoi," P–W, 19 (1937): 833–37, for Crates.

T. L. Heath, *Aristarchus of Samos*, for Heraclides and Aristarchus.

———— *History of Greek Mathematics*, for Eudemus and Eratosthenes.

Hipparchus, *In Arati et Eudoxi Phaenomena commentariorum libri III*; ed. by K. Manitius.

E. Honigmann, "Strabo (Eratosthenes)," P–W, Ser. 2, 4 (1932): 132–36.

Knaack, "Eratosthenes," P–W, 6 (1907): 358–88.

"Krates," S–S, Pt. 2, 1 (1911): 208–10.

H. J. Mette, *Sphairopoiia; Untersuchungen zur Kosmologie des Krates von Pergamon*. Munich, 1936.

J. Mogenet, *Autolycus de Pitane; histoire du texte suivie de l'édition critique des traités de la sphère en mouvement et des levers et couchers*. Louvain, 1950.

Müller-Graupa, "Museion," P–W, 16 (1934): 801–21, for Hellenistic museums and research.

O. Regenbogen, "Theophrastus," P–W, *Supplementband*, 7 (1940): 1354–1562, espec. 1535–39.

G. Sarton, "Euclid and His Time," in his *Ancient Science and Modern Civilization*, 3–36.

W. W. Tarn, *Hellenistic Civilization*. 3d ed. London, 1952.

IV

THE POSIDONIAN AGE

1 See J. F. Dobson, "The Posidonius Myth," *Classical Quarterly*, 12 (1918): 179–95; F. E. Robbins, "Posidonius and the Sources of Pythagorean Arithmology," *Classical Philology*, 15 (1920): 309–22; M. Croiset, in his review of Reinhardt's *Poseidonios*, in *Journal des savants* (1922), 145–52; P. Boyancé, *Études sur le Songe de Scipion* (Limoges, 1936), 87.

2 See *The Geographical Fragments of Hipparchus*, ed. by Dicks, 150–52.

3 Compare, for example, the severe estimate of George Sarton, who had less regard for Posidonius because of his Platonist tendencies and his interest in astrology. See his *History of Science in the Last Three Centuries B.C.*, 255–56.

4 K. Reinhardt, *Poseidonios* (Munich, 1921), 178–83; E. H. Berger, *Die geographischen Fragmente der Eratosthenes*, 109 [n].

5 For a detailed discussion of the literature on Geminus' dates and the character of his writings, see K. Manitius' edition of Geminus' *Elementa astronomiae* (Leipzig, 1898), 237–52, and C. Tittel's article on Geminus in P–W, 7 (1912): 1026–50.

6 P. Duhem, *Le système du monde*, 2 (1914): 157; G. Sarton, *Introduction to the History of Science*, 1:211–12; J. L. E. Dreyer, *History of the Planetary Systems*, 150; and T. L. Heath, *History of Greek Mathematics*, 2:235–36, favor the earlier date. S–S, Pt. 2, 2 (1913): 724; J. O. Thomson, *History of Ancient Geography* (Cambridge, 1948), 215; and W. D. Ross, in his article "Cleomedes," in *The Oxford Classical Dictionary*, favor the second century after Christ.

7 See I. Thomas, *Selections Illustrating the History of Greek Mathematics* (London, 1941) 2:399; T. L. Heath, *Aristarchus of Samos*, 80.

8 See F. E. Robbins, "The Tradition of Greek Arithmology," *Classical Philology*, 16 (1921): 97–123; "Posidonius and the Sources of Greek Arithmology," op. cit., 15 (1920): 309–22; W. H. Roscher, *Die Hebdomadenlehren der griechischen Philosophen und Ärzte* (Leipzig, 1906); *Die hippokratische Schrift von der Siebenzahl in ihrer vierfachen Überlieferung* (Paderborn, 1913).

9 Notably Karl Reinhardt and the scholars who have accepted his view. See his *Poseidonios*, 416, and his article on Posidonius in P–W, 22 (1953): 569. For opposing views see A. E. Taylor, *A Commentary on Plato's Timaeus*, 35–36; L. Edelstein, "The Philosophical System of Posidonius," *American Journal of Philology*, 57 (1936): 304[n].

10 J. L. Heiberg, *Mathematical and Physical Science*, 88, and G. Sarton, *Introduction to the History of Science*, 1:272, assume a single Theon. T. H. Martin, in his edition of Theon's *Liber de astronomia* (Paris, 1849),

8–9, denies the identity on the ground that Theon of Smyrna was not an observer.
11 In Nicomachus of Gerasa, *Introduction to Arithmetic,* chap. 3.
12 Theon of Smyrna, *Expositio rerum mathematicarum ad legendum Platonem utilium;* ed. by E. Hiller(Leipzig, 1878), 141.
13 E. Pais, "The Time and Place in Which Strabo Composed His Geography," *Ancient Italy* (London, 1908), 379–428.

BIBLIOGRAPHY

E. H. Berger, *Geschichte der wissenschaftlichen Erdkunde der Griechen* (2d ed. Leipzig, 1903), for Polybius.
E. H. Bunbury, *A History of Ancient Geography.* Vol. 2 for Polybius and Posidonius.
W. Capelle, "Die griechische Erdkunde und Posidonius," *Neue Jahrbücher für das klassische Altertum, Geschichte und deutsche Literatur,* 45 (1920): 305–24.
————— "Gezeiten," P–W, *Supplementband,* 7 (1940): 208–20.
Cleomedes, *De motu circulari corporum caelestium libri II;* ed. by H. Ziegler. Leipzig, 1891.
Cleomedes, *Die Kreisbewegung der Gestirne;* trans. from the Greek, with notes, by A. Czwalina. Leipzig, 1927.
D. R. Dicks, *The Geographical Fragments of Hipparchus;* ed. by Dicks, for Posidonius and Strabo.
————— "Strabo and the KΛIMATA," *Classical Quarterly,* 50 (1956): 243–47.
J. F. Dobson, "The Posidonius Myth," *Classical Quarterly,* 12 (1918): 179–95.
J. L. E. Dreyer, *History of the Planetary Systems,* for Geminus and Theon.
L. Edelstein, "The Philosophical System of Posidonius," *American Journal of Philology,* 57 (1936): 286–325.
K. von Fritz, "Theon," P–W, Ser. 2, 5 (1934): 2067–75.
Geminus, *Elementa astronomiae;* ed. by K. Manitius. Leipzig, 1898.
F. Gisinger, "Geographie (Polybius)," P–W, *Supplementband,* 4 (1924: 622–29.
————— "Geographie (Poseidonios)," ibid., 630–38.
————— "Geographie (Strabo)," ibid., 638–44.
T. L. Heath, *Aristarchus of Samos,* for Geminus, Cleomedes, and Theon.
————— *History of Greek Mathematics.* Vol. 2 for Posidonius, Geminus, Cleomedes, and Theon.
E. Hiller, "De Adrasti Peripatetici in Platonis Timaeum commentario," *Rheinisches Museum für Philologie N.F.,* 26 (1871): 582–89.
E. Honigmann, "Strabon (Poseidonios)," P–W, Ser. 2, 4 (1932): 109–23.
F. Hultsch, *Poseidonios über die Grösse und Entfernung der Sonne.* Berlin, 1897.

R. M. Jones, "Posidonius and Solar Eschatology," *Classical Philology*, 27 (1932): 113–35.

A. Klotz, *Cäserstudien nebst einer Analyse der strabonischen Beschreibung von Gallien und Britannien.* Leipzig and Berlin, 1910.

A. D. Nock, "Posidonius," *Journal of Roman Studies*, 49 (1959): 1–15.

M. Pohlenz, *Die Stoa; Geschichte einer geistigen Bewegung.* Göttingen, 1955–59.

"Poseidonios," S–S, Pt. 2, 1 (1920): 347–55.

A. Rehm, "Kleomedes," P–W, 11 (1922): 679–94.

K. Reinhardt, *Poseidonios.* Munich, 1921.

———— "Poseidonios von Apameia," P–W, 22 (1953): 558–826.

M. C. P. Schmidt, "Was schrieb Geminos?" *Philologus*, 45 (1886): 63–81.

———— "Zur Isagoge des Geminus," op. cit., 278–320.

J. G. Smyly, "Notes on Theon of Smyrna," *Hermathena*, 14 (1907): 261–79.

Strabo, *Geographica*; ed. by A. Meineke. Leipzig, 1915–25. 3 vols.

———— *The Geography of Strabo*; English trans. by H. L. Jones. London, 1917–32. 8 vols.

"Strabon," S–S, Pt. 2, 1 (1920): 409–15.

Strabon's Geographika in 17 Büchern; ed., with trans. and notes, by W. Aly. Bonn, 1957. 4 vols.

A. E. Taylor, *A Commentary on Plato's Timaeus*, for Posidonius and Theon.

C. Tittel, "Geminos," P–W, 7 (1912): 1026–50.

V

LATE REPUBLICAN TIMES

1 The realization has been growing that the Greeks were not so uninterested in technology as was supposed a generation or two ago. See L. Edelstein, "Recent Trends in the Interpretation of Ancient Science," *Journal of the History of Ideas*, 13 (1952): 579.

2 J. Partsch, *Die Darstellung Europas in dem geographischen Werke des Agrippa* (Breslau, 1875), 80, as cited in Hugo Berger, *Geschichte der wissenschaftlichen Erdkunde der Griechen* (Leipzig, 1893), 4:30.

3 See Moses Hadas, *Ancilla to Classical Reading* (New York, 1954), 16, and the article "Tachygraphy" in *The Oxford Classical Dictionary.*

4 See Pierre Courcelle, *Les lettres grecques en Occident de Macrobe à Cassiodore* (Paris, 1943), 25; F. E. Robbins, "Posidonius and the Sources of Pythagorean Arithmology," *Classical Philology*, 15 (1920): 309–22.

5 "De M. Terentii Varronis Disciplinarum libris commentarius," in his *Kleine philologische Schriften (Opuscula philologica)*, 3 (Leipzig, 1877): 352–402.

6 *Latin Literature* (New York, 1923), 71.

7 *Commentary on Cicero's Dream of Scipio* I.7.1.

BIBLIOGRAPHY

G. L. Beede, *Virgil and Aratus; a Study in the Art of Translation.* Chicago, 1936. A University of Chicago doctoral dissertation.

G. Boissier, *Étude sur la vie et les ouvrages de M. T. Varron.* Paris, 1861.

Cato, *De agricultura;* Varro, *De re rustica;* trans. by W. D. Hooper and H. B. Ash. Cambridge, Mass., 1934.

"M. Porcius Cato," S–H, 1 (1927): 178–93.

F. della Corte, *Varrone; il terzo gran lume romano.* Genoa [1954].

H. Dahlmann, "M. Terentius Varro," P–W, *Supplementband,* 6 (1935): 1172–1277, espec. 1255–59.

M. Fuhrmann, *Das systematische Lehrbuch,* for Cato and Varro.

"General View," T–S, 1 (1891): 87–97, for the encyclopedic interests of the Romans.

W. E. Gillespie, *Vergil, Aratus and Others; the Weather-Sign as a Literary Subject.* Princeton, 1938. A Princeton University doctoral dissertation.

Lucretius, *De rerum natura libri VI;* ed., with intro., critical apparatus, trans., and commentary, by C. Bailey. Oxford, 1947. 3 vols.

"T. Lucretius Carus," S–H, 1 (1927): 271–84.

H. I. Marrou, *A History of Education.*

E. Reitzenstein, *Theophrast bei Epikur und Lucrez.* Heidelberg, 1924.

G. Sarton, *A History of Science in the Last Three Centuries B.C.* (Cambridge, Mass., 1959), for Cato and Varro.

"M. Terentius Varro," S–H, 1 (1927): 555–78, espec. 567–69.

VI

THE AUGUSTAN AGE

1   Both factors are given weight by J. W. Duff, *A Literary History of Rome in the Silver Age* (New York, C. Scribner's Sons, 1933), 127–28.

2   The common practice of Roman geographers of accounting for their lack of interest in remote lands by calling the place names too barbarous to pronounce may have originated with Polybius. In *Histories* III.36 he explains that he does not intend to trace the course of Hannibal's march into Italy by referring to place names which, for his readers, would be unintelligible sounds. Instead he offers a concise geographical conspectus of the known world, setting the demarcations between the continents and giving the general location of countries, large rivers, and seas—the sort of conspectus that became a standard feature of the popular handbooks. For the influence that Polybius exerted upon subsequent geographical writers, Greek and Roman, see E. H. Bunbury, *A History of Ancient Geography,* 2:16–42.

3   A suggestion of Bunbury, 2:358.

4   G. Sarton, *Introduction to the History of Science,* 1:224, and S–H 2

(1935): 387–88, refer to studies that represent Vitruvius as a medieval impostor; J. L. E. Dreyer, *A History of the Planetary Systems*, 128, accepts the view of him as a compiler living c. A.D. 400; J. L. Heiberg, *Mathematics and Physical Science*, 82, doubts that he was a master-builder under Augustus.

5   An example is M. S. Briggs in his chapter on "Building Construction" in *A History of Technology*, ed. by Charles Singer and others, 2:397.

6   See Sarton's article on acoustic vases in his *Introduction to the History of Science*, 3 (1948): 1569–70.

7   See Charles Singer in his chapter on "Science" in *The Legacy of Rome*, ed. by Cyril Bailey, 298. Also T. L. Heath, *A History of Greek Mathematics*, 2:302.

8   Otto Neugebauer considers the frequent occurrence in astrological texts of the 8th degree of Aries as vernal point to be a strong argument for the late date for the introduction of astrology into Greece. See *The Exact Sciences in Antiquity* (2d ed. 1957), 188.

9   Singer, in *The Legacy of Rome*, 284, remarks that Celsus' work is "in many ways the most readable and well-arranged ancient medical work that we have." Heiberg, *Mathematics and Physical Science*, 83, calls it "far and away the best book in the whole range of Roman scientific literature."

10   Owsei Temkin points out Wellman's change of mind from Cassius to Menecrates in his "Celsus' 'On Medicine' and the Ancient Medical Sects," *Bulletin of the History of Medicine, Johns Hopkins University*, 3 (1935): 254.

11   *A. Cornelii Celsi quae supersunt*; ed. by F. Marx (Leipzig and Berlin, 1915), lxxiv.

12   In the article cited above (n. 10), 249–54.

13   *Institutio oratoriae* XII.11.24.

14   *Epistulae* III.5.

15   The English translation of the abstract, prepared from the German version by Cyril Bailey, now appears as an appendix in the third volume of his 1947 edition of Lucretius' *De rerum natura*.

16   See Piero Treves' article "Meteorology," *Oxford Classical Dictionary*. Also T–S, 2 (1892): 46.

BIBLIOGRAPHY

T. C. Allbutt, *Greek Medicine in Rome*. London, 1921.

A. Boëthius, "Vitruvius and the Roman Architecture of His Age," in ΔΡΑΓΜΑ *Martino P. Nilsson A.D. IV ID. IUL. Anno MCMXXXIX dedicatus* (Acta Instituti Romani Regni Sueciae, Series altera, 1939), 1:114–43.

E. H. Bunbury, *A History of Ancient Geography*. Vol. 2 for Mela and for Augustan geography.

Aurelius Cornelius Celsus, *De medicina*; with an English trans. by W. G. Spencer. Cambridge, Mass., 1935–38. 3 vols.

"A. Cornelius Celsus," S–H, 2 (1935), 722–27.

J. Clarke, *Physical Science in the Time of Nero; Being a Translation of the Quaestiones Naturales of Seneca;* by J. Clarke, with notes on the treatise by A. Geike. London, 1910.

F. H. Cramer, *Astrology in Roman Law and Politics* (Philadelphia, 1954), for Seneca.

J. W. Duff, *A Literary History of Rome in the Silver Age,* for Seneca.

M. Fuhrmann, *Das systematische Lehrbuch,* for Vitruvius and Celsus.

F. Gisinger, "Geographie (Agrippa)," P–W, *Supplementband,* 4 (1924): 644–47.

———— "Geographie (Mela)," ibid., 673–75.

———— "Periplus," P–W, 19 (1937): 841–50.

———— "Pomponius Mela," P–W, 21 (1952): 2360–2411.

J. Ilberg, "A. Cornelius Celsus und die Medizin in Rom," *Neue Jahrbücher für das klassische Altertum, Geschichte und deutsche Literatur,* 19 (1907): 378–412.

W. Kubitschek, "Karten (Agrippa)," P–W, 10 (1919): 2100–11.

"Mela," S–H, 2 (1935): 654–56.

Pomponius Mela, *De situ orbis,* In *Collection des auteurs latins;* published under the direction of M. Nisard. Paris, 1883.

K. Miller, *Itineraria romana; römische Reisewege an der Hand der Tabula Peutingeriana.* Stuttgart, 1916.

"L. Annaeus Seneca," S–H, 2 (1935): 679–722, espec. 698–703, 720–22.

J. O. Thomson, *History of Ancient Geography,* for Mela.

H. F. Tozer, *A History of Ancient Geography,* for Agrippa and for the Peutinger Table.

Vitruvius. Pollio, *The Ten Books of Architecture;* trans. by M. H. Morgan. Cambridge, Mass., 1914; reprinted New York, 1960.

Vitruvius, *On Architecture;* ed. and trans. by F. Granger. London, 1931–34. 2 vols.

"Der Baumeister Vitruvius Pollio," S–H, 2 (1935): 386–95.

M. Wellman, *A. Cornelius Celsus; eine Quellenuntersuchung.* Berlin, 1913.

VII

PLINY'S THEORETICAL SCIENCE

1   J. Wight Duff, *A Literary History of Rome in the Silver Age,* 369.
2   The two views are presented in S–H, 2 (1935): 772–73.
3   *The Elder Pliny's Chapters on Chemical Subjects* (London, 1929–32), 2:6–7.
4   The Pliny–Dioscorides–Sextius Niger question has engaged the attention of scholars for more than a century and is still alive. See T–S, 2 (1892): 99; and Pline l'Ancien, *Histoire naturelle, livre XXVII* (Budé edition; Paris, 1959), Intro. by A. Ernout, 7–11.

5 W. Kroll, in *Die Kosmologie des Plinius* (Breslau, 1930) and in his article "Plinius der Ältere" in P–W, 21 (1951): 301–2, has been the leading
proponent of this theory. But see the correctives and cautions introduced by Jean Beaujeu in his Budé edition of Pline l'Ancien, *Histoire
naturelle, livre II* (Paris, 1950), xvi–xviii.

6 On the currency given to astrological doctrines by Posidonius, see P.
Schnabel, *Berossos und die babylonisch-hellenistische Literatur* (Leipzig, 1923), 101–33.

7 Paul Tannery, *Recherches sur l'histoire de l'astronomie ancienne* (Paris,
1893), 325.

8 See Wilfred H. Schoff, *Parthian Stations by Isidore of Charax*; Greek
text, with a trans. and commentary. Philadelphia, 1914.

9 H. E. Wedeck, "The Catalogue in Late and Medieval Latin Poetry,"
Medievalia et Humanistica, 13 (1960): 4–5, sees in such exhaustive
lists as those of place names a device developed in the *suasoria* and
controversia training in the Roman schools of rhetoric.

BIBLIOGRAPHY

E. Bickel, "Neupythagoreische Kosmologie bei den Römern, zu Manilius und
Plinius nat. hist." *Philologus*, 79 (1924): 355–69.

E. H. Bunbury, *A History of Ancient Geography.* Vol. 2.

D. C. Campbell, *C. Plini Secundi Naturalis Historiae Liber Secundus*; notes
on the text, with introductory remarks. Aberdeen, 1942.

D. Detlefsen, *Die Anordnung der geographischen Bücher des Plinius und
ihre Quellen.* Berlin, 1909.

J. W. Duff, *A Literary History of Rome in the Silver Age.*

F. Gisinger, "Geographie (Plinius)," P–W, *Supplementband*, 4 (1924):
675–78.

E. Hoffmann, "Das Proömium zu Plinius' Naturalis historia," *Sokrates N.F.*, 9
(1921): 58–62.

E. Honigmann, *Die sieben Klimata und die* ΠΟΛΕΙΣ ΕΠΙΣΗΜΟΙ; *eine
Untersuchung zur Geschichte der Geographie und Astrologie im Altertum und Mittelalter.* Heidelberg, 1929.

W. Kroll, *Die Kosmologie des Plinius; mit zwei Exkursen von H. Vogt.* Breslau, 1930.

W. Kroll, "Plinius und die Chaldäer," *Hermes*, 65 (1930): 1–13.

W. Kroll and others, "Plinius der Ältere," P–W, 21 (1951): 271–439, espec.
301–7.

Pline l'Ancien, *Histoire naturelle, livre I; livre II*; ed., with trans. and commentary, by Jean Beaujeu; intro. by A. Ernout. Budé edition. Paris,
1950. 2 vols.

Plinius, *Naturalis historia*; ed. by K. Mayhoff. Leipzig, 1892–1909. 6 vols.

"Der Enzyklopädist C. Plinius Secundus," S–H, 2 (1935): 768–83.

Pliny, *Natural History*; ed. and trans. by H. Rackam and W. H. S. Jones.
London, 1938–56. 9 vols.

A. B. West, "Notes on the Multiplication of Cities in Ancient Geography," *Classical Philology*, 18 (1923): 48–67.

VIII

THE SECOND CENTURY

1   *Trionfo della fama* III.38.
2   W. Capelle, "Die Schrift von der Welt," *Neue Jahrbücher für das klassische Altertum, Geschichte und deutsche Literatur*, 8 (1905): 529–68; J. P. Maguire, "The Sources of Ps.–Aristotle De Mundo," *Yale Classical Studies*, 6 (1939): 111–67.
3   Some therapeutic prescriptions of Galen were transmitted to the West and were incorporated in Latin medical compilations, such as the *De medicina* of Cassius Felix (d. A.D. 447). There were professors of medicine who lectured on Galen at Ravenna. Some works or portions of medical writings of Hippocrates, Dioscorides, Soranus, Oribasius, and others were translated into Latin between the fifth and eighth centuries. See H. E. Sigerist, "Medieval Medicine," *University of Pennsylvania Bicentennial Conference; Studies in the History of Science* (Philadelphia, 1941), 47; reprinted in *Toward Modern Science*, ed. by R. M. Palter, 192; G. Sarton, *Introduction to the History of Science*, 1:392, 434. For practical reasons there was a continuing interest in medicine and veterinary medicine in the Latin world from earliest times.
4   See, for example, G. Sarton, *Ancient Science and Modern Civilization*, 50; J. L. E. Dreyer, *History of the Planetary Systems*, 202; C. Singer, *Greek Biology and Greek Medicine* (Oxford, 1922), 122.
5   In Nicomachus of Gerasa, *Introduction to Arithmetic*, chap. 3. Nicomachus' treatise has been exhaustively analyzed there by F. E. Robbins and L. C. Karpinski.
6   Robbins' suggestion, op. cit., 44[n], 185[n].
7   Ibid., 225[n].

BIBLIOGRAPHY

H. Cherniss, "Notes on Plutarch's De Facie in Orbe Lunae," *Classical Philology*, 46 (1951): 137–58.
K. von Fritz, "Theon," P–W, Ser. 2, 5 (1934): 2067–75.
Aulus Gellius, *The Attic Nights*; ed., with an English trans., by J. C. Rolfe (London, 1927–28). 3 vols.
T. L. Heath, *A History of Greek Mathematics*, for Nicomachus and Theon.
E. Hiller, "De Adrasti Peripatetici in Platonis Timaeum commentario," *Rheinisches Museum für Philologie N.F.*, 26 (1871): 582–89.
Pseudo-Iamblichus, *Theologoumena arithmeticae*; ed. by V. de Falco. Leipzig, 1922.

G. Johnson, *The Arithmetical Philosophy of Nicomachus of Gerasa*. Philadelphia, 1916. A University of Pennsylvania doctoral dissertation.

R. M. Jones, *The Platonism of Plutarch*. Menasha, Wis., 1916. A University of Chicago doctoral dissertation.

S. Miller, *Das Verhältnis von Apuleius De mundo zu seiner Vorlage*. Leipzig, 1939; *Philologus Supplementband*, 32, Heft 2.

Nicomachus Gerasenus, *Introductionis arithmeticae libri II*; ed. by R. Hoche. Leipzig, 1866.

Claudius Ptolemaeus, *Handbuch der Astronomie*; trans. from the Greek, with notes, by K. Manitius. Leipzig, 1912–13. 2 vols.

—— *Syntaxis mathematica*; ed. by J. L. Heiberg. Leipzig, 1898–1903. 2 vols.

G. Sarton, *Galen of Pergamon*. Lawrence, Kans., 1954.

—— "Ptolemy and His Time," in *Ancient Science and Modern Civilization*, 37–73.

J. G. Smyly, "Notes on Theon of Smyrna," *Hermathena*, 14 (1907): 261–79.

Theo Smyrnaeus, *Expositio rerum mathematicarum ad legendum Platonem utilium*; ed. by E. Hiller. Leipzig, 1878.

—— *Liber de astronomia cum Sereni fragmento*; ed. with Latin trans. and notes, by T. H. Martin. Paris, 1849.

IX

THIRD- AND FOURTH-CENTURY COSMOGRAPHY

1 Theodor Mommsen himself was influenced by such arguments and was inclined to place Solinus' date before Constantine. See his edition of Solinus' *Collectanea rerum memorabilium* vi. Mommsen's opinion has been widely adopted by scholars.

2 Ibid., xvii–xix.

3 Golding's version was reproduced in 1955 in a facsimile edition, with a brief introduction by George Kish, of the University of Michigan.

4 This is the attitude of Pierre Duhem, *Le système du monde*, 2:418–19. Most scholars regard Chalcidius as a Christian.

5 In his edition of Theon of Smyrna's *Liber de astronomia*, 18.

6 For some of these arguments see E. Hiller, "De Adrasti Peripatetici in Platonis Timaeum commentario," *Rheinisches Museum für Philologie* N.F., 26 (1871): 582–84.

7 Plutarch, *De procreatione animae* 1027d, is the source of this information.

8 *De divisione naturae* III.27 (Migne, *Patrologia latina*, 122:697–98).

9 For opinions about the importance of Chalcidius' commentary in transmitting Platonic ideas to the Latin West during the Middle Ages, see R. Klibansky, *The Continuity of the Platonic Tradition* (London, 1939), 27–28; A. E. Taylor, *Plato; the Man and His Work* (6th ed. New York,

1956), 437. A new edition of Chalcidius' work, edited by J. H. Waszink, is announced for publication in 1962. It will be Volume 4 in the series *Corpus Platonicum medii aevi*, edited by R. Klibansky. The book will discuss the sources and influence of Chalcidius and the filiation of manuscripts, which number about 150.

BIBLIOGRAPHY

C. R. Beazley, *The Dawn of Modern Geography*. London, 1897–1906. Vol. 1 for Solinus.
E. H. Bunbury, *A History of Ancient Geography*. Vol. 2 for Solinus.
Chalcidius, *Platonis Timaeus interprete Chalcidio cum eiusdem commentario*; ed. by J. Wrobel. Leipzig, 1876.
"Chalcidius," S–H, 4 (1914): 137–39.
Diehl, "Iulius Solinus," P–W, 10 (1917): 823–38.
P. Duhem, *Le système du monde*. Vols. 2 and 3 for Chalcidius.
R. M. Jones, "Chalcidius and Neo-Platonism," *Classical Philology*, 13 (1918): 194–208.
W. Kroll, "Chalcidius," P–W, 3 (1897): 2042–43.
M. L. W. Laistner, "The Decay of Geographical Knowledge and the Decline of Exploration," in *Travel and Travellers of the Middle Ages*, ed. by A. P. Newton (London, 1930), 19–38.
C. Iulius Solinus, *Collectanea rerum memorabilium*; ed. by T. Mommsen. Berlin, 1895; reprinted Berlin, 1958.
"C. Julius Solinus," S–H, 3 (1922): 224–27.
"C. Julius Solinus," T–H, 2 (1892): 291–93.
W. H. Stahl, "Dominant Traditions in Early Medieval Latin Science," *Isis*, 50 (1959): 95–124.
B. W. Switalski, *Des Chalcidius Kommentar zu Platos Timaeus; eine historisch-kritische Untersuchung*. Münster, 1902.
A. E. Taylor, *A Commentary on Plato's Timaeus*, for Chalcidius.

X

FIFTH-CENTURY COMMENTATOR

1 J. E. Sandys, *A History of Classical Scholarship* (Cambridge, Eng., 1906), 1:238, and T. R. Glover, *Life and Letters in the Fourth Century* (Cambridge, Eng., 1901), 172, are influenced by these considerations.
2 G. Wissowa, *De Macrobii Saturnaliorum fontibus capita tria; dissertatio inauguralis philologica* (Breslau, 1880), 15.
3 Duhem, *Le système du monde*, 2:482, feels certain about this.
4 For a comparison of the texts see W. H. Roscher, *Die hippokratische Schrift von der Siebenzahl* (Paderborn, 1913), 92–97.

5   See P. Courcelle, *Les lettres grecques en Occident*, 20–35, 394–95; W. H. Stahl, *Macrobius' Commentary on the Dream of Scipio* (New York, 1952), 23–39.

6   J. L. E. Dreyer, *History of the Planetary Systems*, 129–30; T. L. Heath, *Aristarchus of Samos*, 258–59; Duhem, *Le système du monde*, 3:51–52.

7   R. M. Jones, "Posidonius and Solar Eschatology," *Classical Philology*, 27 (1932): 122.

8   F. Hultsch, *Poseidonios über die Grösse und Entfernung der Sonne* (Berlin, 1897), 5–6, gives serious consideration to Macrobius' statement, an idle discussion in the light of the fact that Macrobius knew as much about Eratosthenes as a modern schoolboy knows about Isaac Newton. Eratosthenes' attempt to measure the dimensions of the earth and sun was a familiar tag about him in the early Middle Ages just as the story of the apple is the inevitable tag about Newton for persons who know little else about him.

9   See M. C. Andrews, "The Study and Classification of Medieval Mappae Mundi," *Archaeologia*, 75 (1925): 61–76; C. R. Beazley, *The Dawn of Modern Geography*, 2:573–76, 625–26; J. O. Thomson, *History of Ancient Geography*, 203; J. K. Wright, *The Geographical Lore of the Time of the Crusades* (New York, 1925), 18–19, 56, 158–59.

10   Lactantius, *Divinae institutiones* III.24; St. Augustine, *De civitate Dei* XVI.9.

11   J. K. Wright, *Geographical Lore*, 56–57, 160–61, 383–84, 386; M. L. W. Laistner, *Thought and Letters in Western Europe A.D. 500–900* (New York, 1931), 145–46; Dreyer, *History of the Planetary Systems*, 224; Thomson, *History of Ancient Geography*, 386.

12   See K. Mras, "Macrobius' Kommentar zu Ciceros Somnium," *Sitzungsberichte der preussischen Akademie der Wissenschaften, Berlin, philosophisch-historische Klasse* (1933), 266–67; Stahl, *Macrobius' Commentary on the Dream of Scipio*, 194.

BIBLIOGRAPHY

P. Boyancé, *Études sur le Songe de Scipion*. Limoges, 1936.

S. T. Collins, *The Interpretation of Vergil, with Special Reference to Macrobius*. Oxford, 1909.

P. Courcelle, *Les lettres grecques en Occident*.

S. Dill, *Roman Society in the Last Century of the Western Empire*. London, 2d ed., 1905; reprinted New York, 1958.

P. Duhem, *Le système du monde*. Vol. 3 for Macrobius.

F. Gisinger, "Perioikoi," P–W, 19 (1937): 833–37.

T. R. Glover, *Life and Letters in the Fourth Century*. Cambridge, 1901.

T. L. Heath, *Aristarchus of Samos*.

P. Henry, *Plotin et l'occident; Firmicus Maternus, Marius Victorinus, Saint Augustin et Macrobe*. Louvain, 1934.

Macrobius, *Opera quae supersunt*; ed. by L. von Jan. Leipzig and Quedlin-
berg, 1848–52. 2 vols.

Macrobius; ed. by F. Eyssenhardt. Leipzig, 1893.

"Ambrosius Macrobius Theodosius," S–H, 4 (1920): 189–96.

P. M. Schedler, *Die Philosophie des Macrobius und ihr Einfluss auf die Wissenschaft des christlichen Mittelalters*. Münster, 1916.

P. Wessner, "Macrobius," P–W, 14 (1930): 170–98.

T. Whittaker, *Macrobius; or Philosophy, Science and Letters in the Year 400*.
Cambridge, 1923.

XI

VARRONIAN ENCYCLOPEDIST

1 This last was the view of Eyssenhardt, a 19th-century editor of Capella's
work, and was accepted by the latest editor, A. Dick, in his (Leipzig,
1925) edition, xxv.

2 The documentation for much of this discussion of Capella's geography
and astronomy may be found in my preliminary paper, "Dominant Traditions in Early Medieval Latin Science," *Isis*, 50 (1959): 98–111.

3 J. K. Wright, *Geographical Lore*, 55, 155; G. H. T. Kimble, *Geography
in the Middle Ages* (London, 1936), 8–9, 24.

4 Capella's summary remark (VI.650) covers the omission of Pliny
III.101–150. For some examples of the garbling or misinterpreting of
Pliny's geographical account, compare Capella VI.594 and Pliny II.180
(Pliny's phrase *serius nobis illi* becomes a fictitious Servius Nobilis in
Capella); Capella 621 and Pliny II.170 (the story of the Indian captives
sailing past Germany); Capella 632 and Pliny III.16–18 (Capella inserts two sentences about Tarraconensis into a passage on Baetica);
Capella 643 and Pliny III.77 (Pliny notes that the Greek name for
the Balearic Isles is Gymnasiae; Capella makes the Gymnasiae separate
islands); Capella 649 and Pliny III.94–95 (Capella confuses the Auson-
ian Sea and the first gulf of Europe); Capella 650 and Pliny III.97 (by
fusing two sentences, Capella makes Croton 85 miles distant from the
Acroceraunian Promontory); Capella 651 and Pliny IV.1 (Pliny is speaking about the legendary lore of Greece; Capella omits the sentence but
tacks Greece onto a list of place names); Capella 654 and Pliny IV.32
(Capella gives the same figures for the extent of Thessaly that Pliny gives
for Epirus, Achaea, Attica, and Thessaly); Capella 656 gives the height
of Mt. Haemus as six miles; Pliny IV.41 says the distance to the summit
is six miles; Capella 659, while discussing Euboea, gives the length of
Boeotia; Pliny IV.63 says that Euboea stretches along the entire length
of Boeotia, and gives the length of Euboea; Capella 665 and Pliny
VI.34–35 (Pliny deals with the Arimphaei on his southern periplus; Ca-

pella, who is actually following Solinus' *Collectanea* 17.2 here, includes them in his northern periplus); Pliny VI.84–88 tells the story of a visit of Ceylonese ambassadors to Rome and of their surprise to find that in northern latitudes the sun rises on the left as one faces south; Capella 697 omits the story and has the sun rising on the left in Ceylon.

5 E. R. Curtius, *European Literature and the Latin Middle Ages* (New York, 1953), 104, finds in Capella's allegorical female figures a rhetorical cliché for the supernaturally endowed young-old woman; a more familiar example is recognizable in the matron Philosophy who appears to Boethius in his *Consolation of Philosophy.*

6 By Paul Tannery and T. L. Heath. See Heath's *Aristarchus of Samos,* 259.

BIBLIOGRAPHY

J. W. H. Atkins, *English Literary Criticism; the Medieval Phase.* Cambridge and New York, 1943.

C. R. Beazley, *The Dawn of Modern Geography.* Vol. 1 for Capella.

Martianus Mineius Felix Capella, *De nuptiis Philologiae et Mercurri et de septem artibus liberalibus libri IX;* ed. by U. F. Kopp. Frankfurt, 1836.

Martianus Capella; ed. by A. Dick. Leipzig, 1925.

"Martianus Capella," S–H, 4 (1920): 166–70.

"Martianus Capella," T–S, 2 (1892): 446–49.

P. R. Cole, *Later Roman Education in Ausonius, Capella and the Theodosian Code;* with trans. and commentary. New York, 1909. A Columbia University Teachers' College doctoral dissertation.

P. Courcelle, *Les lettres grecques en Occident.*

P. Duhem, *Le système du monde.* Vol. 3.

Dunchad, *Glossae in Martianum;* ed. by C. E. Lutz. Lancaster, Pa., 1944; American Philological Association Monographs, No. 12.

Johannes Scottus [Erigena], *Annotationes in Marcianum;* ed. by C. E. Lutz. Cambridge, Mass., 1939.

F. Gisinger, "Perioikoi," P–W, 19 (1937): 833–37.

T. Haarhoff, *Schools of Gaul; a Study of Pagan and Christian Education in the Last Century of the Western Empire.* London, 1920.

M. L. W. Laistner, "Martianus Capella and His Ninth Century Commentators," *Bulletin of the John Rylands Library,* 9 (1925): 130–38.

C. E. Lutz, "The Commentary of Remigius of Auxerre on Martianus Capella," *Medieval Studies,* 19 (1957): 137–56.

Nicomachus of Gerasa, *Introduction to Arithmetic;* trans. by M. L. D'Ooge.

J. G. Préaux, "Le commentaire de Martin de Laon sur l'oeuvre de Martianus Capella," *Latomus,* 12 (1953): 437–59.

P. Wessner, "Martianus Capella," P–W, 14 (1930): 2003–16.

J. K. Wright, *Geographical Lore.*

XII

CLASSICAL LEARNING UNDER THE OSTROGOTHS

1   *Decline and Fall,* chap. 30; ed. by J. B. Bury (10th ed. London, 1930), 4:197–98.
2   Ibid.
3   See W. Bark, "Theodoric vs. Boethius: Vindication and Apology," *American Historical Review,* 49 (1944): 410–26.
4   *Vorlesungen über Geschichte der Mathematik* (Leipzig, 1880), 1:490–99.
5   In M. L. D'Ooge's translation of Nicomachus' *Introduction,* 132–37.
6   See J. F. Mountford, "Greek Music and Its Relation to Modern Times," *Journal of Hellenic Studies,* 40 (1920): 39–40.
7   This is the opinion of T. L. Heath, *History of Greek Mathematics,* 1:359; S–H, 4, Pt. 2 (1920): 153.
8   *Epistulae* 8 (Migne, *Patrologia latina,* 139:203).
9   H. M. Klinkenberg, "Der Verfall des Quadriviums in frühen Mittelalter," in *Artes liberales; von der antiken Bildung zur Wissenschaft des Mittelalters,* ed. by Josef Koch (Leiden and Cologne, 1959), 1–32, espec. 1–2, 8–20, sees a fundamental break in the traditional conception of the quadrivium occurring between Boethius and Cassiodorus. Boethius deals with the quadrivium subjects according to Pythagorean concepts of a knowable universe based upon numbers and proportions; Cassiodorus does not admit the possibility of such an approach to the wonders of God's mind. Cassiodorus copies a passage from Boethius—while attributing the views to "the ancients" (*veteres*)—to the effect that astronomy enables us to "reason about the heavens, to investigate the nature of the celestial machine, and to draw some conclusions about the secrets veiled in the immensity of the universe." The last portion of Cassiodorus' statement suppresses a key phrase in the original and departs from the position of Boethius, who says that we are enabled "to come to know the Creator of the universe." The two statements follow: *totamque illam machinam supernam indagabili ratione discutere et inspectiva mentis sublimitate ex aliqua parte colligere quod tantae magnitudinis arcana velaverunt* (Cassiodorus); *totamque illam machinam supernam indagabili ratione aliter discutere et inspectiva mentis sublimitate ex aliqua parte colligere et agnoscere mundi factorem, qui tanta et talia arcana velavit* (Boethius).
10   *Les lettres grecques en Occident,* 330–32.
11   See L. W. Jones' translation of Cassiodorus' *Introduction to Divine and Human Readings* (New York, 1946), 22–23.

BIBLIOGRAPHY

H. M. Barrett, *Boethius; Some Aspects of His Times and Work.* Cambridge, Eng., 1940.

"Anicius Manlius Severinus Boethius," S–H, 4, Pt. 2 (1920): 148–66.

Boetius, *Opera omnia; tomus prior*; in *Patrologiae cursus completus, series latina*, ed. by J. P. Migne (Paris, 1847), Vol. 63.

Cassiodorus Senator, *An Introduction to Divine and Human Readings*; trans. with an intro. and notes, by L. W. Jones. New York, 1946.

────── *Institutiones*; ed. by R. A. B. Mynors. Oxford, 1937.

────── *Variae*; ed. by T. Mommsen. Berlin, 1894.

"Flavius Magnus Aurelius Cassiodorus Senator," S–H, 4, Pt. 2 (1920): 92–109.

P. Courcelle, *Les lettres grecques en Occident*, for Boethius and Cassiodorus.

Y. Courtonne, *Saint Basile et l'héllenisme; étude sur la rencontre de la pensée chrétienne avec la sagesse antique dans l'Hexaméron de Basile le Grand*. Paris, 1934.

E. S. Duckett, *The Gateway to the Middle Ages* (New York, 1938), for Boethius and Cassiodorus.

Hartmann, "Cassiodorus," P–W, 3 (1897): 1671–76.

T. Hodgkin, *The Letters of Cassiodorus*; a condensed trans. of the *Variae* of Cassiodorus Senator, with an intro. London, 1886.

M. L. W. Laistner, *Thought and Letters in Western Europe A.D. 500–900* (2d ed. Ithaca, N.Y., 1957), for Boethius and Cassiodorus.

P. Lehmann, "Cassiodorstudien," *Philologus* (1913–17), 71:278–79; 72:503–17; 73:253–73; 74:351–83.

M. Manitius, "Boethius," in his *Geschichte der lateinischen Literatur des Mittelalters* (Munich, 1911), 1:22–36.

────── "Cassiodorus Senator," op. cit., 36–52.

H. I. Marrou, *Saint Augustin et la fin de la culture antique* (Paris, 1938), for Cassiodorus.

A. Momigliano, "Cassiodorus and the Italian Culture of His Time," *Proceedings of the British Academy*, 41 (1955): 207–45.

O. Paul, *Boetius und die griechische Harmonik; des Anicius Manlius Severinus Boetius fünf Bücher über die Musik*; trans. from the Latin, with special attention to the subject of Greek harmony. Leipzig, 1872.

F. E. Robbins, *The Hexaemeral Literature; a Study of the Greek and Latin Commentaries on Genesis*. Chicago, 1912. A University of Chicago doctoral dissertation.

L. Schrade, "Music in the Philosophy of Boethius," *Musical Quarterly*, 33 (1947): 188–200.

H. Thiele, "Cassidor, seine Klostergründung Vivarium und sein Nachwirken im Mittelalter," *Studien und Mitteilungen zur Geschichte des Benediktiner-Ordens und seiner Zweige*, 50 (1932): 378–419.

V. van de Vyver, "Cassiodor et son oeuvre," *Speculum*, 6 (1931): 244–92.

────── "Les Institutiones de Cassiodore et sa fondation à Vivarium," *Revue Bénédictine*, 53 (1941): 59–88.

XIII

THE BORDERLANDS

1   H. Kettner, *Varronische Studien* (Halle, 1875), 2–37; M. Klussmann, *Excerpta Tertulliana in Isidori Hispalensis Etymologiis.* Hamburg, 1892.

2   Cassiodorus Senator, *Institutiones*; ed. by Mynors, 193.

3   For a summary of the sources of Isidore's *Etymologies*, see M. Manitius, *Geschichte der lateinischen Literatur*, 1:62–66. I am indebted to Manitius for the statements here about Festus, Servius, and the early Latin writers.

4   M. C. Andrews, "The Study and Classification of Medieval Mappae Mundi," *Archaeologia*, 75 (1925): 61–76; the article "Map" in the *Encyclopaedia Britannica* (11th ed. Cambridge, Eng., 1911), 17:638; C. R. Beazley, *Dawn of Modern Geography*, 2:576–79, 618–33.

5   Laistner in his chapter, "The Library of the Venerable Bede," in A. H. Thompson, ed., *Bede; Life, Times and Writings* (Oxford, 1935), 237–66, and in his article, "Bede as a Classical and a Patristic Scholar," *Transactions of the Royal Historical Society*, Ser. 4, 16 (1933): 69–94.

6   G. Sarton in his *Introduction to the History of Science*, 1:510, attributes this judgment to Poole. Poole was undoubtedly thinking of the Latin West and the early Middle Ages when he made this estimate of Bede as a chronologer.

7   The sources of Bede's calendar treatises and his close ties with Irish computus literature have been thoroughly investigated by Charles W. Jones. Students of medieval chronology will find his edition of these treatises (Beda, *Opera de temporibus*, Cambridge, Mass., 1943) an indispensable reference work.

8   The mean interval between the time of high water and that of the moon's previous meridian transit.

9   See C. W. Jones, "Bede and Vegetius," *Classical Review*, 46 (1932): 248–49.

BIBLIOGRAPHY

Bede, *The Complete Works of Venerable Bede*; Latin text, Eng. trans. of the historical works, and a life of the author, by J. A. Giles (London, 1843–44). 12 vols.; scientific works in Vol. 6.

Beda, *Opera historica*; ed. by C. Plummer. Oxford, 1896.

C. H. Beeson, *Isidor-Studien.* Munich, 1913.

R. R. Bolgar, *The Classical Heritage and Its Beneficiaries* (Cambridge, Eng., 1954), for Bede.

E. Brehaut, *An Encyclopedist of the Dark Ages; Isidore of Seville.* New York, 1912.

Y. Courtonne, *Saint Basile et l'héllenisme.*

R. Davis, "Bede's Early Reading," *Speculum,* 8 (1933): 179–95.

J. L. E. Dreyer, "Medieval Astronomy," in *Studies in the History and Method of Science,* ed. by Charles Singer (Oxford, 1921), 2:102–20; reprinted in *Toward Modern Science,* ed. by R. M. Palter, 1:235–56.

————— "Medieval Cosmology," in his *History of the Planetary Systems,* 207–39.

E. S. Duckett, *Anglo-Saxon Saints and Scholars* (New York, 1947), for Bede.

P. Duhem, *Le système du monde.* Vol. 3 for Isidore and Bede.

J. Fontaine, *Isidore de Seville et la culture classique dans l'Espagne wisigothique.* Paris, 1959. 2 vols.

Isidorus Hispalensis, *De natura rerum liber;* ed. by G. Becker. Berlin, 1857.

————— *Etymologiarum libri XX;* ed. by F. Lindemann. Leipzig, 1833.

————— *Etymologiarum sive Originum libri XX;* ed. by W. M. Lindsay. Oxford, 1911. 2 vols.

"Isidorus," T–S, 2 (1892): 566–69.

M. L. W. Laistner, *Thought and Letters in Western Europe A.D. 500–900,* for Isidore and Bede.

M. Manitius, "Beda," in his *Geschichte der lateinischen Literatur,* 1:70–87.

————— "Isidor von Sevilla," op.cit., 52–70.

Philipp, "Isidorus von Sevilla (Hispalensis)," P–W, 9 (1916): 2069–80.

G. R. Stephens, *The Knowledge of Greek in England in the Middle Ages.* Philadelphia, 1933. A University of Pennsylvania doctoral dissertation.

A. H. Thompson, ed., *Bede; His Life, Times and Writings.* Oxford, 1935.

K. Werner, *Beda der Ehrwürdige und seine Zeit.* Vienna, rev. ed., 1881.

XIV

ROMAN SURVIVALS

1 See M. L. W. Laistner's opinion about the importance of Traube's contributions in his article, "Bede as a Classical and a Patristic Scholar," *Transactions of the Royal Historical Society,* Ser. 4, 16 (1933): 69–70.

2 See Mommsen's edition of Solinus, *Collectanea,* xxviii.

3 C. R. Beazley, *Dawn of Modern Geography,* 1:162.

4 P. Duhem, *Le système du monde,* 3:62. On Erigena's direct borrowings from Macrobius, see C. E. Lutz' edition of Johannes Scottus, *Annotationes in Marcianum* xx.

5 *De divisione naturae* III.33 (Migne, *Patrologia latina,* 122:716–18).

6 *Le système du monde,* 3:71–76.

7 Ibid., 62.

8 Ibid., 76–87.

9 On Lupitus and the book on astrology, see ibid., 164–65; M. Manitius, *Geschichte der lateinischen Literatur,* 2 (1923): 730; J. M. Millás Vallicrosa, *Nuevos estudios sobre historia de la ciencia española* (Barcelona, 1960), 2:97–98.

10   *Le système du monde,* 3:164–65.
11   Ibid., 165.
12   J. K. Wright, *Geographical Lore,* 103.
13   Ibid., 158, and for an illustration of the map, 122.
14   *Le système du monde,* 3:193.
15   Ibid., 129.
16   P. M. Schedler, *Die Philosophie des Macrobius,* 144–46.
17   *Le système du monde,* 3:130–52.
18   In his article "Du quadrivium à la physique," in *Artes liberales; von der antiken Bildung zur Wissenchaft des Mittelalters,* ed. by Josef Koch, 112.

BIBLIOGRAPHY

C. R. Beazley, *Dawn of Modern Geography.* Vols. 1 and 2 for Dicuil and Lambert of St. Omer.
C. H. Beeson, *Isidor-Studien,* for Isidore influence outside of Spain.
H. Bett, *Johannes Scotus Erigena; a Study in Medieval Philosophy.* Cambridge, Eng., 1925.
Cassiodorus Senator, *An Introduction to Divine and Human Readings,* trans. by L. W. Jones, for Hrabanus' borrowing from Cassiodorus.
A. C. Crombie, *Medieval and Early Modern Science.* New York, 1959. 2 vols.
Dicuil, *Liber de mensura orbis terrae;* ed. by C. A. Walckenaer. Paris, 1807.
J. L. E. Dreyer ,"Medieval Astronomy," in *Studies in the History and Method of Science,* ed. by Charles Singer, 2:102–20, for Erigena.
——— "Medieval Cosmology," in his *History of the Planetary Systems,* 207–39.
E. S. Duckett, *Alcuin, Friend of Charlemagne; His World and His Work.* New York, 1951.
P. Duhem, *Le système du monde.* Vol. 3 for Erigena, Helpericus, Pseudo-Bede, Gerbert, William of Conches, and Thierry.
Einhard, *Life of Charlemagne;* the Latin text, ed. with intro. and notes by H. W. Garrod and R. B. Mowat. Oxford, 1915.
E. von Erhardt-Siebold and R. von Erhardt, *The Astronomy of Johannes Scotus Erigena.* Baltimore, 1940.
——— *Cosmology in the Annotationes in Marcianum; More Light on Erigena's Astronomy.* Baltimore, 1940.
Johannes Scottus, [Erigena], *Annotationes in Marcianum;* ed. by C. E. Lutz. Cambridge, Mass. 1939.
Euclid, *The Thirteen Books of Euclid's Elements;* trans., with an intro. and commentary, by T. L. Heath. Vol. 1 for Euclid's tradition in the Middle Ages.
E. Garin, *Studi sul platonismo medievale* (Florence, 1958), for Manegold of Lautenbach, Pseudo-Bede, Bernard of Chartres, and William of Conches.

F. Gisinger, "Oikumene (Kirchenväter)," P–W, 17 (1937): 2162–64.

C. H. Haskins, *The Renaissance of the Twelfth Century*. Cambridge, Mass., 1927; reprinted New York, 1957.

—— *Studies in the History of Mediaeval Science*. Cambridge, Mass., 1924.

Helpericus, *Liber de computo*; in *Patrologia latina*, ed. by J. P. Migne (Paris, 1879), Vol. 137, cols. 15–48.

E. Honigmann, *Die sieben Klimata und die* ΠΟΛΕΙΣ ΕΠΙΣΗΜΟΙ, for traditions of Eratosthenes and Hipparchus in the Middle Ages.

Hugh of St. Victor, *The Didascalicon; a Medieval Guide to the Arts*. Trans. from the Latin with an intro. and notes, by Jerome Taylor. New York, 1961.

W. P. Ker, *The Dark Ages*. Edinburgh, 1904.

M. L. W. Laistner, *Thought and Letters in Western Europe A.D. 500–900*.

Lambert of St. Omer, *Liber floridus*; in *Patrologia latina*, ed. by J. P. Migne (Paris, 1854), Vol. 163, cols. 1003–32.

H. P. Lattin, "The Eleventh Century MS Munich 14436; its contribution to the history of coordinates, of logic, of German studies in France," *Isis*, 38 (1948): 205–25.

G. Leff, *Medieval Thought from Saint Augustine to Ockham*. Harmondsworth, Eng., 1958.

C. E. Lutz, "The Commentary of Remigius of Auxerre on Martianus Capella," *Mediaeval Studies*, 19 (1957): 137–56.

M. Manitius, *Geschichte der lateinischen Literatur des Mittelalters* (Munich, 1911–31) for nearly all authors; the chief work in the field. 3 vols.

E. Norden, "Die Stellung des Artes liberales im mittelalterlichen Bildungswesen," in *Die antike Kunstprosa* (Leipzig, 1898), 2:670–87.

K. Rück, "Die Naturalis historia des Plinius im Mittelalter," *Sitzungsberichte der Bayerischen Akademie der Wissenschaften, Munich, philosophisch-historische Classe*, 1 (1898): 203–318.

P. M. Schedler, *Die Philosophie des Macrobius und ihr Einfluss auf die Wissenschaft des christlichen Mittelalters* (Münster, 1916), for Pseudo-Bede, Erigena, Helpericus, William of Conches, and Bartholomew.

G. E. SeBoyar, "Bartholomaeus Anglicus and His Encyclopaedia," *Journal of English and Germanic Philology*, 19 (1920): 168–89.

E. M. Sanford, "The Use of Classical Latin Authors in the *Libri Manuales*," *Transactions of the American Philological Association*, 55 (1924): 190–248.

P. Tannery, *Sciences exactes au moyen âge* (Toulouse and Paris, 1922), for Pseudo-Boethius and Hugh of St. Victor.

L. Thorndike, *The Sphere of Sacrobosco and Its Commentators*. Chicago, 1949.

L. Traube, *Vorlesungen und Abhandlungen*; Vol. 2: *Einleitung in die lateinische Philologie des Mittelalters*. Munich, 1911.

Überweg-Heinze-Geyer, *Grundriss der Geschichte der Philosophie* (11th ed. Basle, 1955). Vol. 2.

L. Wallach, *Alcuin and Charlemagne; Studies in Carolingian History and Literature*. Ithaca, N.Y., 1959.

J. K. Wright, "Notes on the Knowledge of Latitudes and Longitudes in the Middle Ages," *Isis*, 5 (1923): 75–98.

XV

CONCLUSION

1   A. C. Crombie, *Medieval and Early Modern Science*, New York, 1959.

2   Otto Neugebauer also sees a connection between the popularization of science and the teaching profession in his article "Exact Science in Antiquity," *University of Pennsylvania Bicentennial Conference: Studies in Civilization* (Philadelphia, 1941), 26; reprinted in *Toward Modern Science*, ed. by R. M. Palter, 17–18.

3   On the dreariness of school books and recitations, see H. I. Marrou, *History of Education in Antiquity*, 165–69, 279–81; E. B. Castle, *Ancient Education and Today* (Harmondsworth, Eng., 1961), 127–29. On the standardization of educational practices, see Marrou, Pt. 2, chap. 2 and Pt. 3, chap. 2.

4   In the beautiful passage in which Virgil expresses his aspiration to compose a poem on cosmography (*Georgics* II.475–92), he indicates in lines 483–84 his own misgivings about his ability to comprehend such subjects.

5   *Policraticus* 471c, crediting Pythagoras and Plotinus but actually drawing upon Cicero, *De senectute* 73, and Macrobius, *Commentary on the Dream of Scipio* I.13.9–20.

INDEX